T0362164

A First Course *in* Optimization

A First Course *in* Optimization

Charles L. Byrne

University of Massachusetts Lowell
Lowell, USA

CRC Press is an imprint of the
Taylor & Francis Group, an **informa** business
A CHAPMAN & HALL BOOK

CRC Press
Taylor & Francis Group
6000 Broken Sound Parkway NW, Suite 300
Boca Raton, FL 33487-2742

First issued in hardback 2019

ISBN-13: 978-1-4822-2656-0 (hbk)

Library of Congress Cataloging-in-Publication Data

Byrne, Charles L., 1947- author.
A first course in optimization / Charles L. Byrne, Department of Mathematical Sciences, University of Massachusetts Lowell.
pages cm
Summary: "Designed for graduate students and advanced undergraduate students, this text provides a much-needed contemporary introduction to optimization. Emphasizing general problems and the underlying theory, it covers the fundamental problems of constrained and unconstrained optimization, linear and convex programming, fundamental iterative solution algorithms, gradient methods, the Newton-Raphson algorithm and its variants, and sequential unconstrained optimization methods. The book presents the necessary mathematical tools and results as well as applications, such as game theory"-- Provided by publisher.
Includes bibliographical references and index.
ISBN 978-1-4822-2656-0 (hardback : acid-free paper) 1. Mathematical optimization--Textbooks. I. Title.

QA402.5.B97 2014
519.6--dc23 2014016437

Visit the Taylor & Francis Web site at
http://www.taylorandfrancis.com

and the CRC Press Web site at
http://www.crcpress.com

To my wife Eileen,
with thanks for forty-four wonderful years of marriage.

Contents

Preface

This book originated as a set of notes I used for a one-semester course in optimization taken by advanced undergraduate and beginning graduate students in the mathematical sciences and engineering. For the past several years I have used versions of this book as the text for that course. In that course, the focus is on generality, with emphasis on the fundamental problems of constrained and unconstrained optimization, linear and convex programming, on fundamental iterative solution algorithms, such as gradient methods, the Newton-Raphson algorithm and its variants, and more general iterative optimization methods, and on the necessary mathematical tools and results that provide the proper foundation for our discussions. I include some applications, such as game theory, but the emphasis is on general problems and the underlying theory. As with most introductory mathematics courses, the course has both an explicit and an implicit objective. Explicitly, I want the student to learn the basics of continuous optimization. Implicitly, I want the student to understand better the mathematics that he or she has already been exposed to in previous classes.

One reason for the usefulness of optimization in applied mathematics is that Nature herself often optimizes, or perhaps better, Nature economizes. The patterns and various sizes of tree branches form efficient communication networks; the hexagonal structures in honeycombs are an efficient way to fill the space; the shape of a soap bubble minimizes the potential energy in the surface tension; and so on. Optimization means maximizing or minimizing some function of one or, more often, several variables. The function to be optimized is called the *objective function*. There are two distinct types of applications that lead to optimization problems, which, to give them names, I shall call *natural* problems of optimization and *problems of inference*.

Natural problems of optimization are ones in which optimizing a given function is, more or less, the sole and natural objective. The main goal, maximum profits, shortest commute, is not open to question, although the precise function involved will depend on the simplifications adopted as the real-world problem is turned into mathematics. Examples of such problems are a manufacturer seeking to maximize profits, subject to whatever restrictions the situation imposes, or a commuter trying to minimize the time it takes to get to work, subject, of course, to speed limits. In converting the real-world problem to a mathematical problem, the manufacturer may or may not ignore non-linearities such as economies of scale, and the commuter may or may not employ probabilistic models of traffic density. The resulting mathematical optimization problem to be solved will depend on such choices, but the original real-world problem is one of optimization, nevertheless.

In addition to natural optimization problems, there are what we might call *artificial* optimization problems, often *problems of inference*, for which optimization provides useful tools, but is not the primary objective. Such problems often arise in solving under-determined or over-determined systems of linear equations, in statistical parameter estimation, and many other places. As we shall see, in both types of problems, the optimization usually cannot be performed by algebraic means alone and iterative algorithms are required.

Operations Research (OR) is a broad field involving a variety of applied optimization problems. Wars and organized violence have always given impetus to technological advances, most significantly during the twentieth century. An important step was taken when scientists employed by the military realized that studying and improving the use of existing technology could be as important as inventing new technology. Conducting research into on-going operations, that is, doing operations research, led to the search for better, indeed, optimal, ways to schedule ships entering port, to design convoys, to paint the under-sides of aircraft, to hunt submarines, and many other seemingly mundane tasks [138]. Problems having to do with the allocation of limited resources arise in a wide variety of applications, all of which fall under the broad umbrella of OR.

Sometimes we may want to optimize more than one thing; that is, we may have more than one objective function that we wish to optimize. In image processing, we may want to find an image as close as possible to measured data, but one that also has sharp edges. In general, such multiple-objective optimization is not possible; what is best in one respect need not be best in other respects. In such cases, it is common to create a single objective function that is a combination, a sum perhaps, of the original objective functions, and then to optimize this combined objective function. In this way, the optimizer of the combined objective function provides a sort of compromise.

The goal of simultaneously optimizing more than one objective function, the so-called *multiple-objective-function problem*, is a common feature of many economics problems, such as bargaining situations, in which the various parties all wish to steer the outcome to their own advantage. Typically, of course, no single solution will optimize everyone's objective function. Bargaining is then a method for finding a solution that, in some sense, makes everyone equally happy or unhappy. A *Nash equilibrium* is such a solution.

In 1994, the mathematician John Nash was awarded the Nobel Prize in Economics for his work in optimization and mathematical economics. His theory of equilibria is fundamental in the study of bargaining and game theory. In her book *A Beautiful Mind* [163], later made into a movie of the same name starring Russell Crowe, Sylvia Nasar tells the touching story of Nash's struggle with schizophrenia, said to have been made more acute by his obsession with the mysteries of quantum mechanics. Strictly speaking, there is no Nobel Prize in Economics; what he received is "The Central Bank of Sweden Prize in Economic Science in Memory of Alfred Nobel," which was instituted seventy years after Nobel created his prizes. Nevertheless, it is commonly spoken of as a Nobel Prize.

Problems of inference often involve estimates to be made from measured data. Such problems arise in many remote-sensing applications, radio astronomy, or medical imaging, for example, in which, for practical reasons, the data obtained are insufficient or too noisy to specify a unique source, and one turns to optimization methods, such as likelihood maximization or least-squares, to provide usable approximations. In such cases, it is not the optimization of a function that concerns us, but the optimization of technique. We cannot know which reconstructed image is the best, in the sense of most closely describing the true situation, but we do know which techniques of reconstruction are "best" in some specific sense. We choose techniques such as likelihood or entropy maximization, or least-mean-squares minimization, because these methods are "optimal" in some sense, not because any single result obtained using these methods is guaranteed to be the best. Generally, these methods are "best" in some average sense; indeed, this is the basic idea in statistical estimation.

The mathematical tools required do not usually depend on which type of problem we are trying to solve. A manufacturer may use the theory of linear programming to maximize profits, while an oncologist may use likelihood maximization to image a tumor and linear programming to determine a suitable spatial distribution of radiation intensities for the therapy. The only difference, perhaps, is that the doctor may have some choice in how, or even whether or not, to involve optimization in solving the medical problems, while the manufacturer's problem is an optimization problem from the start, and a linear programming problem once the mathematical model is selected.

The optimization problems we shall discuss differ from one another in the nature of the functions being optimized and in the constraints that may or may not be imposed. The constraints themselves may involve other functions; we may wish to minimize $f(x)$, subject to the constraint $g(x) \leq 0$. The functions may or may not be differentiable. They may or may not be linear. If they are not linear, they may be convex. They may become linear or convex once we change variables. The various problem types have names, such as Linear Programming, Quadratic Programming, Geometric

Programming, and Convex Programming; the use of the term "programming" is an historical accident and has no connection with computer programming.

In many of the problems we shall consider, finding exact or approximate solutions of systems of linear equations plays a central role. When an exact solution is sought and the number of equations and the number of unknowns are small, methods such as Gauss elimination can be used. However, in applications such as medical imaging it is common to encounter problems involving hundreds or even thousands of equations and unknowns. It is also common to prefer inexact solutions to exact ones, when the equations involve noisy, measured data. Even when the number of equations and unknowns is large, there may not be enough data to specify a unique solution, and we need to incorporate prior knowledge about the desired answer. Such is the case with medical tomographic imaging, in which the images are artificially discretized approximations of parts of the interior of the body.

For problems involving many variables, it is important to use algorithms that provide an acceptable approximation of the solution in a reasonable amount of time. For medical tomography image reconstruction in a clinical setting, the algorithm must reconstruct a useful image from scanning data in the time it takes for the next patient to be scanned, which is roughly fifteen minutes. Some of the algorithms we shall encounter work fine on small problems, but require far too much time when the problem is large. Figuring out ways to speed up convergence is an important part of iterative optimization.

As we noted earlier, optimization is often used when the data pertaining to a desired mathematical object (a function, a vectorized image, etc.) is not sufficient to specify uniquely one solution to the problem. It is common in remote-sensing problems for there to be more than one mathematical solution that fits the measured data. In such cases, it is helpful to turn to optimization, and seek the solution consistent with the data that is closest to what we expect the correct answer to look like. This means that we must somehow incorporate prior knowledge about the desired answer into the algorithm for finding it.

In this book we only scratch the surface of optimization; we ignore entire branches of optimization, such as discrete optimization, combinatorial optimization, stochastic optimization, and many others. The companion volume [64] continues the discussion of continuous optimization, focusing on the use of iterative optimization methods in inverse problems. I have posted copies of most of my articles referenced in the bibliography on my website, http://faculty.uml.edu/cbyrne/cbyrne.html.

Overview

Chapter 1: Optimization Without Calculus

Although optimization is a central topic in applied mathematics, most of us first encountered this subject in Calculus I, as an application of differentiation. I was surprised to learn how much could be done without calculus, relying only on a handful of inequalities. The purpose of this chapter is to present optimization in a way we all could have learned it in elementary and high school, but didn't. The key topics in this chapter are the Arithmetic-Geometric Mean Inequality and Cauchy's Inequality.

Chapter 2: Geometric Programming

Although Geometric Programming (GP) is a fairly specialized topic, a discussion of the GP problem is a quite appropriate place to begin. This chapter on the GP problem depends heavily on the Arithmetic-Geometric Mean Inequality discussed in the previous chapter, while introducing new themes, such as duality, primal and dual problems, and iterative computation, that will be revisited several times throughout the course.

Chapter 3: Basic Analysis

Here we review basic notions from analysis, such as limits of sequences in $\mathbb{R}^J$, continuous functions, and completeness. Less familiar topics that play important roles in optimization, such as semi-continuity, are also discussed.

Chapter 4: Convex Sets

One of the fundamental problems in continuous optimization, perhaps the fundamental problem, is to minimize a real-valued function of several real variables over a subset of $\mathbb{R}^J$. In order to obtain a satisfactory theory we need to impose certain restrictions on the functions and on the subsets; convexity is perhaps the most general condition that still permits the development of an adequate theory. In this chapter we discuss convex sets, leaving the subject of convex functions to a subsequent chapter. Theorems of the Alternative, which we discuss here, play a major role in the duality theory of linear programming.

Chapter 5: Vector Spaces and Matrices

Convex sets defined by linear equations and inequalities play a major role in optimization, particularly in linear programming, and matrix algebra is therefore an important tool in these cases. In this chapter we present a short summary of the basic notions of matrix theory and linear algebra.

Chapter 6: Linear Programming

Linear Programming (LP) problems are the most important of all the optimization problems, and the most tractable. These problems arise in a wide variety of applications, and efficient algorithms for solving LP problems, such as Dantzig's Simplex Method, are among the most frequently used routines in computational mathematics. In this chapter we see once again the notion of duality that we will first encounter in Chapter 2.

Chapter 7: Matrix Games and Optimization

Two-person zero-sum matrix games provide a nice illustration of the techniques of linear programming. In this chapter we use tools from LP to establish John von Neumann's Fundamental Theorem of Game Theory.

Chapter 8: Differentiation

While the concepts of directional derivatives and gradients are familiar enough, they are not the whole story of differentiation. In this chapter we consider the Gâteaux derivative and the Fréchet derivative, along with several examples. This chapter can be skipped without harm to the reader.

Chapter 9: Convex Functions

In this chapter we review the basic calculus of real-valued functions of one and several real variables, with emphasis on convex functions, in preparation for the study of convex programming and iterative optimization.

Chapter 10: Convex Programming

Convex programming involves the minimization of convex functions, subject to convex constraints. This is perhaps the most general class of continuous optimization problems for which a fairly complete theory exists. Once again, duality plays an important role. Some of the discussion here concerning Lagrange multipliers should be familiar to students.

Chapter 11: Iterative Optimization

In iterative methods, we begin with a chosen vector and perform some operation to get the next vector. The same operation is then performed again to get the third vector, and so on. The goal is to generate a sequence of vectors that converges to the solution of the problem. Such iterative methods are needed when the original problem has no algebraic solution, such as finding $\sqrt{3}$, and also when the problem involves too many variables to make an algebraic approach feasible, such as solving a large system of linear equations. In this chapter we consider the application of iterative methods to the problem of minimizing a function of several variables.

Chapter 12: Solving Systems of Linear Equations

This chapter is a sequel to the previous one, in the sense that here we focus on the use of iterative methods to solve large systems of linear equations. Specialized algorithms for incorporating positivity constraints are also considered.

Chapter 13: Conjugate-Direction Methods

The problem here is to find a least-squares solution of a large system of linear equations. The conjugate-gradient method (CGM) is tailored to this specific problem, although extensions of this method have been used for more general optimization. In theory, the CGM converges to a solution in a finite number of steps, but in practice, the CGM is viewed as an iterative method.

Chapter 14: Operators

In this chapter we consider several classes of linear and nonlinear operators that play important roles in optimization.

Chapter 15: Looking Ahead

We have just scratched the surface of optimization. In this chapter we preview sequential unconstrained iterative optimization methods, discussed in greater detail in the companion volume [64].

Chapter 1

Optimization Without Calculus

1.1 Chapter Summary

In our study of optimization, we shall encounter a number of sophisticated techniques, involving first and second partial derivatives, systems of linear equations, nonlinear operators, specialized distance measures, and so on. It is good to begin by looking at what can be accomplished without sophisticated techniques, even without calculus. It is possible to achieve much with powerful, yet simple, inequalities. Someone once remarked, exaggerating slightly, that, in the right hands, the Cauchy Inequality and integration by parts are all that are really needed. Some of the discussion in this chapter follows that in Niven [168].

Students typically encounter optimization problems as applications of differentiation, while the possibility of optimizing without calculus is left unexplored. In this chapter we develop the Arithmetic Mean-Geometric Mean Inequality, abbreviated the AGM Inequality, from the convexity of the logarithm function, use the AGM to derive several important inequalities, including Cauchy's Inequality, and then discuss optimization methods based on the Arithmetic Mean-Geometric Mean Inequality and Cauchy's Inequality.

1.2 The Arithmetic Mean-Geometric Mean Inequality

Let $x_1, ..., x_N$ be positive numbers. According to the famous *Arithmetic Mean-Geometric Mean Inequality*, abbreviated AGM Inequality,

$$G = (x_1 \cdot x_2 \cdots x_N)^{1/N} \leq A = \frac{1}{N}(x_1 + x_2 + ... + x_N), \tag{1.1}$$

with equality if and only if $x_1 = x_2 = ... = x_N$. To prove this, consider the following modification of the product $x_1 \cdots x_N$. Replace the smallest of the x_n, call it x, with A and the largest, call it y, with $x + y - A$. This modification does not change the arithmetic mean of the N numbers, but the product increases, unless $x = y = A$ already, since $xy \leq A(x + y - A)$. (Why?) We repeat this modification, until all the x_n approach A, at which point the product reaches its maximum.

For example, $2 \cdot 3 \cdot 4 \cdot 6 \cdot 20$ becomes $3 \cdot 4 \cdot 6 \cdot 7 \cdot 15$, and then $4 \cdot 6 \cdot 7 \cdot 7 \cdot 11$, $6 \cdot 7 \cdot 7 \cdot 7 \cdot 8$, and finally $7 \cdot 7 \cdot 7 \cdot 7 \cdot 7$.

1.3 Applying the AGM Inequality: the Number e

We can use the AGM Inequality to show that

$$\lim_{n \to \infty} \left(1 + \frac{1}{n}\right)^n = e.$$

Let $f(n) = \left(1 + \frac{1}{n}\right)^n$, the product of the $n+1$ numbers $1, 1 + \frac{1}{n}, ..., 1 + \frac{1}{n}$. Applying the AGM Inequality, we obtain the inequality

$$f(n) \leq \left(\frac{n+2}{n+1}\right)^{n+1} = f(n+1),$$

so we know that the sequence $\{f(n)\}$ is increasing. Now define $g(n) = (1 + \frac{1}{n})^{n+1}$; we show that $g(n) \leq g(n-1)$ and $f(n) \leq g(m)$, for all positive integers m and n. Consider $(1-\frac{1}{n})^n$, the product of the $n+1$ numbers $1, 1-\frac{1}{n}, ..., 1-\frac{1}{n}$. Applying the AGM Inequality, we find that

$$\Big(1-\frac{1}{n+1}\Big)^{n+1} \geq \Big(1-\frac{1}{n}\Big)^n,$$

or

$$\Big(\frac{n}{n+1}\Big)^{n+1} \geq \Big(\frac{n-1}{n}\Big)^n.$$

Taking reciprocals, we get $g(n) \leq g(n-1)$. Since $f(n) < g(n)$ and $\{f(n)\}$ is increasing, while $\{g(n)\}$ is decreasing, we can conclude that $f(n) \leq g(m)$, for all positive integers m and n. Both sequences therefore have limits. Because the difference

$$g(n) - f(n) = \frac{1}{n}\Big(1+\frac{1}{n}\Big)^n \to 0,$$

as $n \to \infty$, we conclude that the limits are the same. This common limit we can define as the number e.

1.4 Extending the AGM Inequality

We begin with the notion of a *convex function* of a real variable.

Definition 1.1 *A function $f : \mathbb{R} \to \mathbb{R}$ is said to be* convex *over an interval (a, b) if*

$$f(a_1t_1 + a_2t_2 + ... + a_Nt_N) \leq a_1f(t_1) + a_2f(t_2) + ... + a_Nf(t_N),$$

for all positive integers N, all positive real numbers a_n that sum to one, and all real numbers t_n in (a, b).

Suppose, once again, that $x_1, ..., x_N$ are positive numbers. Let $a_1, ..., a_N$ be positive real numbers that sum to one. Then the *Generalized AGM Inequality* (GAGM Inequality) is

$$x_1^{a_1}x_2^{a_2}\cdots x_N^{a_N} \leq a_1x_1 + a_2x_2 + ... + a_Nx_N, \tag{1.2}$$

with equality if and only if $x_1 = x_2 = ... = x_N$. We can prove this using the convexity of the function $f(x) = -\log x$. The inequality (1.2) generalizes the one in (1.1).

If a function $f(x)$ is twice differentiable on (a, b), then $f(x)$ is convex over (a, b) if and only if the second derivative of $f(x)$ is nonnegative on (a, b). For example, the function $f(x) = -\log x$ is convex on the positive x-axis. The GAGM Inequality follows immediately.

1.5 Optimization Using the AGM Inequality

We illustrate the use of the AGM Inequality for optimization through several examples.

1.5.1 Example 1: Minimize This Sum

Find the minimum of the function

$$f(x,y) = \frac{12}{x} + \frac{18}{y} + xy,$$

over positive x and y.

We note that the three terms in the sum have a fixed product of 216, so, by the AGM Inequality, the smallest value of $\frac{1}{3}f(x,y)$ is $(216)^{1/3} = 6$. The smallest value occurs when the three terms are equal. Therefore, each is equal to 6; so $x = 2$ and $y = 3$. The smallest value of $f(x,y)$ is therefore 18.

1.5.2 Example 2: Maximize This Product

Find the maximum value of the product

$$f(x,y) = xy(72 - 3x - 4y),$$

over positive x and y.

The terms x, y and $72 - 3x - 4y$ do not have a constant sum, but the terms $3x$, $4y$ and $72 - 3x - 4y$ do have a constant sum, namely 72, so we rewrite $f(x,y)$ as

$$f(x,y) = \frac{1}{12}(3x)(4y)(72 - 3x - 4y).$$

By the AGM Inequality, the product $(3x)(4y)(72 - 3x - 4y)$ is maximized when the factors $3x$, $4y$ and $72 - 3x - 4y$ are each equal to 24, so when $x = 8$ and $y = 6$. The maximum value of the product is then 1152.

1.5.3 Example 3: A Harder Problem?

Both of the previous two problems can be solved using the standard calculus technique of setting the two first partial derivatives to zero. Here is an example that may not be so easily solved in that way: minimize the function

$$f(x,y) = 4x + \frac{x}{y^2} + \frac{4y}{x},$$

over positive values of x and y. Try taking the first partial derivatives and setting them both to zero. Even if we manage to solve this system of coupled nonlinear equations, deciding if we actually have found the minimum may not be easy; we would have to investigate the second derivative matrix, the Hessian matrix. We can employ the AGM Inequality by rewriting $f(x, y)$ as

$$f(x, y) = 4\Big(\frac{4x + \frac{x}{y^2} + \frac{2y}{x} + \frac{2y}{x}}{4}\Big).$$

The product of the four terms in the arithmetic mean expression is 16, so the GM is 2. Therefore, $\frac{1}{4}f(x, y) \geq 2$, with equality when all four terms are equal to 2; that is, $4x = 2$, so that $x = \frac{1}{2}$ and $\frac{2y}{x} = 2$, so $y = \frac{1}{2}$ also. The minimum value of $f(x, y)$ is then 8.

1.6 The Hölder and Minkowski Inequalities

Let $c = (c_1, ..., c_N)$ and $d = (d_1, ..., d_N)$ be vectors with complex entries and let p and q be positive real numbers such that

$$\frac{1}{p} + \frac{1}{q} = 1.$$

The p-norm of c is defined to be

$$\|c\|_p = \left(\sum_{n=1}^{N} |c_n|^p\right)^{1/p},$$

with the q-norm of d, denoted $\|d\|_q$, defined similarly.

1.6.1 Hölder's Inequality

Hölder's Inequality is the following:

$$\sum_{n=1}^{N} |c_n d_n| \leq \|c\|_p \|d\|_q,$$

with equality if and only if

$$\left(\frac{|c_n|}{\|c\|_p}\right)^p = \left(\frac{|d_n|}{\|d\|_q}\right)^q$$

for each n.

Hölder's Inequality follows from the GAGM Inequality. To see this, we fix n and apply Inequality (1.2), with

$$x_1 = \left(\frac{|c_n|}{\|c\|_p}\right)^p,$$

$$a_1 = \frac{1}{p},$$

$$x_2 = \left(\frac{|d_n|}{\|d\|_q}\right)^q,$$

and

$$a_2 = \frac{1}{q}.$$

From (1.2) we then have

$$\left(\frac{|c_n|}{\|c\|_p}\right)\left(\frac{|d_n|}{\|d\|_q}\right) \leq \frac{1}{p}\left(\frac{|c_n|}{\|c\|_p}\right)^p + \frac{1}{q}\left(\frac{|d_n|}{\|d\|_q}\right)^q.$$

Now sum both sides over the index n.

1.6.2 Minkowski's Inequality

Minkowski's Inequality, which is a consequence of Hölder's Inequality, states that

$$\|c+d\|_p \leq \|c\|_p + \|d\|_p\,;$$

it is the triangle inequality for the metric induced by the p-norm.

To prove Minkowski's Inequality, we write

$$\sum_{n=1}^{N} |c_n + d_n|^p \leq \sum_{n=1}^{N} |c_n||c_n + d_n|^{p-1} + \sum_{n=1}^{N} |d_n||c_n + d_n|^{p-1}.$$

Then we apply Hölder's Inequality to both of the sums on the right side of the equation.

1.7 Cauchy's Inequality

For the choices $p = q = 2$, Hölder's Inequality becomes the famous Cauchy Inequality, which we rederive in a different way in this section.

For simplicity, we assume now that the vectors have real entries and for notational convenience later we use x_n and y_n in place of c_n and d_n.

Let $x = (x_1, ..., x_N)^T$ and $y = (y_1, ..., y_N)^T$ be column vectors with real entries. The set of all such vectors we denote by $\mathbb{R}^N$; when we allow the entries to be complex we get the set $\mathbb{C}^N$. The *inner product* of x and y is

$$\langle x, y \rangle = x_1 y_1 + x_2 y_2 + ... + x_N y_N = x \cdot y = y^T x.$$

The two-norm of the vector x is

$$\|x\|_2 = \sqrt{\langle x, x \rangle}.$$

Cauchy's Inequality is

$$|\langle x, y \rangle| \leq \|x\|_2 \, \|y\|_2,$$

with equality if and only if there is a real number a such that $x = ay$.

A vector $x = (x_1, ..., x_N)^T$ in the real N-dimensional space $\mathbb{R}^N$ can be viewed in two slightly different ways. The first way is to imagine x as simply a point in that space; for example, if $N = 2$, then $x = (x_1, x_2)$ would be the point in two-dimensional space having x_1 for its first coordinate and x_2 for its second. When we speak of the norm of x, which we think of as a length, we could be thinking of the distance from the origin to the point x. But we could also be thinking of the length of the directed line segment that extends from the origin to the point x. This line segment is also commonly denoted just x. There will be times when we want to think of the members of $\mathbb{R}^N$ as points. At other times, we shall prefer to view them as directed line segments; for example, if x and y are two points in $\mathbb{R}^N$, their difference, $x - y$, is more likely to be viewed as the directed line segment extending from y to x, rather than a third point situated somewhere else in $\mathbb{R}^N$. We shall make no explicit distinction between the two views, but rely on the situation to tell us which one is the better interpretation.

To prove Cauchy's Inequality, we begin with the fact that, for every real number t,

$$0 \leq \|x - ty\|_2^2 = \|x\|_2^2 - (2\langle x, y \rangle)t + \|y\|_2^2 t^2.$$

This quadratic in the variable t is never negative, so cannot have two distinct real roots. It follows that the term under the radical sign in the quadratic equation must be nonpositive, that is,

$$(2\langle x, y \rangle)^2 - 4\|y\|_2^2 \|x\|_2^2 \leq 0. \tag{1.3}$$

We have equality in (1.3) if and only if the quadratic has a double real root, say $t = a$. Then we have

$$\|x - ay\|_2^2 = 0.$$

As an aside, suppose we had allowed the variable t to be complex. Clearly

$\|x - ty\|$ cannot be zero for any non-real value of t. Doesn't this contradict the fact that every quadratic has two roots in the complex plane?

We can interpret Cauchy's Inequality as providing an upper bound for the quantity

$$\left(\sum_{n=1}^{N} x_n y_n\right)^2.$$

The *Pólya–Szegö Inequality* provides a lower bound for the same quantity.

Theorem 1.1 (The Pólya–Szegö Inequality) *Let $0 < m_1 \le x_n \le M_1$ and $0 < m_2 \le y_n \le M_2$, for all n. Then*

$$\sum_{n=1}^{N} x_n^2 \sum_{n=1}^{N} y_n^2 \le \frac{M_1 M_2 + m_1 m_2}{\sqrt{4 m_1 m_2 M_1 M_2}} \left(\sum_{n=1}^{N} x_n y_n\right)^2.$$

1.8 Optimizing Using Cauchy's Inequality

We present three examples to illustrate the use of Cauchy's Inequality in optimization.

1.8.1 Example 4: A Constrained Optimization

Find the largest and smallest values of the function

$$f(x, y, z) = 2x + 3y + 6z,$$

among the points (x, y, z) with $x^2 + y^2 + z^2 = 1$.

From Cauchy's Inequality we know that

$$49 = (2^2 + 3^2 + 6^2)(x^2 + y^2 + z^2) \ge (2x + 3y + 6z)^2,$$

so that $f(x, y, z)$ lies in the interval $[-7, 7]$. We have equality in Cauchy's Inequality if and only if the vector $(2, 3, 6)$ is parallel to the vector (x, y, z), that is

$$\frac{x}{2} = \frac{y}{3} = \frac{z}{6}.$$

It follows that $x = t$, $y = \frac{3}{2}t$, and $z = 3t$, with $t^2 = \frac{4}{49}$. The smallest value of $f(x, y, z)$ is -7, when $x = -\frac{2}{7}$, and the largest value is $+7$, when $x = \frac{2}{7}$.

1.8.2 Example 5: A Basic Estimation Problem

The simplest problem in estimation theory is to estimate the value of a constant c, given J data values $z_j = c + v_j$, $j = 1, ..., J$, where the v_j are random variables representing additive noise or measurement error. Assume that the expected values of the v_j are $E(v_j) = 0$, the v_j are uncorrelated, so $E(v_j v_k) = 0$ for j different from k, and the variances of the v_j are $E(v_j^2) = \sigma_j^2 > 0$. A *linear* estimate of c has the form

$$\hat{c} = \sum_{j=1}^{J} b_j z_j.$$

The estimate $\hat{c}$ is *unbiased* if $E(\hat{c}) = c$, which forces $\sum_{j=1}^{J} b_j = 1$. The *best* linear unbiased estimator, the BLUE, is the one for which $E((\hat{c} - c)^2)$ is minimized. This means that the b_j must minimize

$$E\left(\sum_{j=1}^{J}\sum_{k=1}^{J} b_j b_k v_j v_k\right) = \sum_{j=1}^{J} b_j^2 \sigma_j^2,$$

subject to

$$\sum_{j=1}^{J} b_j = 1. \tag{1.4}$$

To solve this minimization problem, we turn to Cauchy's Inequality.

We can write

$$1 = \sum_{j=1}^{J} b_j = \sum_{j=1}^{J} (b_j \sigma_j)\frac{1}{\sigma_j}.$$

Cauchy's Inequality then tells us that

$$1 \leq \sqrt{\sum_{j=1}^{J} b_j^2 \sigma_j^2}\sqrt{\sum_{j=1}^{J} \frac{1}{\sigma_j^2}},$$

with equality if and only if there is a constant, say λ, such that

$$b_j \sigma_j = \lambda \frac{1}{\sigma_j},$$

for each j. So we have

$$b_j = \lambda \frac{1}{\sigma_j^2},$$

for each j. Summing on both sides and using Equation (1.4), we find that

$$\lambda = 1/\sum_{j=1}^{J} \frac{1}{\sigma_j^2}.$$

The BLUE is therefore

$$\hat{c} = \lambda \sum_{j=1}^{J} \frac{z_j}{\sigma_j^2}.$$

When the variances σ_j^2 are all the same, the BLUE is simply the arithmetic mean of the data values z_j.

1.8.3 Example 6: A Filtering Problem

One of the fundamental operations in signal processing is filtering the data vector $x = \gamma s + n$, to remove the noise component n, while leaving the signal component s relatively unaltered [53]. This can be done to estimate γ, the amount of the signal vector s present. It can also be done to detect if the signal is present at all, that is, to decide if $\gamma > 0$. The noise is typically known only through its *covariance matrix* Q, which is the positive-definite, symmetric matrix having for its entries $Q_{jk} = E(n_j n_k)$. The filter usually is linear and takes the form of an estimate of γ:

$$\hat{\gamma} = b^T x,$$

where the superscript T denotes matrix transpose. We want $|b^T s|^2$ large, and, on average, $|b^T n|^2$ small; that is, we want $E(|b^T n|^2) = b^T E(nn^T) b = b^T Q b$ small. The best choice is the vector b that maximizes the *gain* of the filter, that is, the ratio

$$|b^T s|^2 / b^T Q b.$$

We can solve this problem using the Cauchy Inequality.

Definition 1.2 *Let S be a square matrix. A nonzero vector u is an* eigenvector *of S if there is a scalar λ such that $Su = \lambda u$. Then the scalar λ is said to be an* eigenvalue *of S associated with the eigenvector u.*

Definition 1.3 *The* transpose, *$B = A^T$, of an M by N matrix A is the N by M matrix having the entries $B_{n,m} = A_{m,n}$.*

Definition 1.4 *A square matrix S is* symmetric *if $S^T = S$.*

A basic theorem in linear algebra is that, for any symmetric N by N matrix S, $\mathbb{R}^N$ has an orthonormal basis consisting of mutually orthogonal, norm-one eigenvectors of S. We then define U to be the matrix whose columns are these orthonormal eigenvectors u^n and L to be the diagonal matrix with the associated eigenvalues λ_n on the diagonal. We can easily see that U is an *orthogonal matrix*, that is, $U^TU = I$. We can then write

$$S = ULU^T; \tag{1.5}$$

this is the *eigenvalue/eigenvector decomposition* of S. The eigenvalues of a symmetric S are always real numbers.

Definition 1.5 *A J by J symmetric matrix Q is* nonnegative definite *if, for every x in $\mathbb{R}^J$, we have $x^TQx \geq 0$. If $x^TQx > 0$ whenever x is not the zero vector, then Q is said to be* positive definite.

In Exercise 1.13 the reader is asked to show that the eigenvalues of a nonnegative (positive) definite matrix are always nonnegative (positive).

A covariance matrix Q is always nonnegative definite, since

$$x^TQx = E\left(\left(\sum_{j=1}^{J} x_j n_j\right)^2\right).$$

Therefore, its eigenvalues are nonnegative; typically, they are actually positive, as we shall assume now. We then let $C = U\sqrt{L}U^T$, which is called the *symmetric square root* of Q since $Q = C^2 = C^TC$. The Cauchy Inequality then tells us that

$$(b^Ts)^2 = (b^TCC^{-1}s)^2 \leq [b^TCC^Tb][s^T(C^{-1})^TC^{-1}s],$$

with equality if and only if the vectors C^Tb and $C^{-1}s$ are parallel. It follows that

$$b = \alpha(CC^T)^{-1}s = \alpha Q^{-1}s,$$

for any constant α. It is standard practice to select α so that $b^Ts = 1$, therefore $\alpha = 1/s^TQ^{-1}s$ and the optimal filter b is

$$b = \frac{1}{s^TQ^{-1}s}Q^{-1}s.$$

1.9 An Inner Product for Square Matrices

The *trace* of a square matrix M, denoted tr$M$, is the sum of the entries down the main diagonal. Given square matrices A and B with real entries, the trace of the product B^TA defines the *trace inner product*, that is

$$\langle A, B\rangle = \text{tr}(B^TA),$$

where the superscript T denotes the transpose of a matrix. This inner product can then be used to define a norm of A, called the *Frobenius norm*, by

$$\|A\|_F = \sqrt{\langle A, A\rangle} = \sqrt{\text{tr}(A^T A)}.$$

It is easy to see that the Frobenius norm of an M by N matrix A is the two-norm of its vectorization, vec(A), which is the MN by 1 column vector whose entries are those of A.

From the eigenvector/eigenvalue decomposition, we know that, for every symmetric matrix S, there is an orthogonal matrix U such that

$$S = UD(\lambda(S))U^T,$$

where $\lambda(S) = (\lambda_1, ..., \lambda_N)$ is a vector whose entries are eigenvalues of the symmetric matrix S, and $D(\lambda(S))$ is the diagonal matrix whose entries are the entries of $\lambda(S)$. Then we can easily see that

$$\|S\|_F = \|\lambda(S)\|_2.$$

Denote by $[\lambda(S)]$ the vector of eigenvalues of S, ordered in nonincreasing order. We have the following result.

Theorem 1.2 (Fan's Theorem) *Any real symmetric matrices S and R satisfy the inequality*

$$\text{tr}(SR) \leq \langle[\lambda(S)], [\lambda(R)]\rangle,$$

with equality if and only if there is an orthogonal matrix U such that

$$S = UD([\lambda(S)])U^T,$$

and

$$R = UD([\lambda(R)])U^T.$$

From linear algebra, we know that S and R can be simultaneously diagonalized if and only if they commute; this is a stronger condition than simultaneous diagonalization.

If S and R are diagonal matrices already, then Fan's Theorem tells us that

$$\langle\lambda(S), \lambda(R)\rangle \leq \langle[\lambda(S)], [\lambda(R)]\rangle.$$

Since any real vectors x and y are $\lambda(S)$ and $\lambda(R)$, for some symmetric S and R, respectively, we have the following theorem.

Theorem 1.3 (The Hardy–Littlewood–Polya Inequality) *Let x and y be vectors in $\mathbb{R}^N$. Then*

$$\langle x, y\rangle \leq \langle[x], [y]\rangle.$$

Most of the optimization problems discussed in this chapter fall under the heading of Geometric Programming, which we shall present in a more formal way in Chapter 2.

1.10 Discrete Allocation Problems

Most of the optimization problems we consider in this book are *continuous* problems, in the sense that the variables involved are free to take on values within a continuum. A large branch of optimization deals with *discrete* problems; the interested reader should consult [137]. Typically, these discrete problems can be solved, in principle, by an exhaustive checking of a large, but finite, number of possibilities; what is needed is a faster method. The *optimal allocation* problem is a good example of a discrete optimization problem.

We have n different jobs to assign to n different people. For $i = 1, ..., n$ and $j = 1, ..., n$ the quantity C_{ij} is the cost of having person i do job j. The n by n matrix C with these entries is the *cost matrix*. An *assignment* is a selection of n entries of C so that no two are in the same column or the same row; that is, everybody gets one job. Our goal is to find an assignment that minimizes the total cost.

We know that there are $n!$ ways to make assignments, so one solution method would be to determine the cost of each of these assignments and select the cheapest. But for large n this is impractical. We want an algorithm that will solve the problem with less calculation. The algorithm we present here, discovered in the 1930's by two Hungarian mathematicians, is called, unimaginatively, the Hungarian Method.

To illustrate, suppose there are three people and three jobs, and the cost matrix is

$$C = \begin{bmatrix} 53 & 96 & 37 \\ 47 & 87 & 41 \\ 60 & 92 & 36 \end{bmatrix}.$$

The number 41 in the second row, third column indicates that it costs 41 dollars to have the second person perform the third job.

The algorithm is as follows:

Step 1: Subtract the minimum of each row from all the entries of that row. This is equivalent to saying that each person charges a minimum amount just to be considered, which must be paid regardless of the allocation made. All we can hope to do now is to reduce the remaining costs. Subtracting these fixed costs, which do not depend on the allocations, does not change the optimal solution.

The new matrix is then

$$\begin{bmatrix} 16 & 59 & 0 \\ 6 & 46 & 0 \\ 24 & 56 & 0 \end{bmatrix}.$$

Step 2: Subtract each column minimum from the entries of its column. This is equivalent to saying that each job has a minimum cost, regardless of who performs it, perhaps for materials, say, or a permit. Subtracting those costs does not change the optimal solution. The matrix becomes

$$\begin{bmatrix} 10 & 13 & 0 \\ 0 & 0 & 0 \\ 18 & 10 & 0 \end{bmatrix}.$$

Step 3: Draw a line through the least number of rows and columns that results in all zeros being covered by a line; here I have put in boldface the entries covered by a line. The matrix becomes

$$\begin{bmatrix} 10 & 13 & \mathbf{0} \\ \mathbf{0} & \mathbf{0} & \mathbf{0} \\ 18 & 10 & \mathbf{0} \end{bmatrix}.$$

We have used a total of two lines, one row and one column. What we are searching for is a set of zeros such that each row and each column contains a zero. Then n lines will be required to cover the zeros.

Step 4: If the number of lines just drawn is n we have finished; the zeros just covered by a line tell us the assignment we want. Since n lines are needed, there must be a zero in each row and in each column. In our example, we are not finished.

Step 5: If, as in our example, the number of lines drawn is fewer than n, determine the smallest entry not yet covered by a line (not boldface, here). It is 10 in our example. Then subtract this number from all the uncovered entries and add it to all the entries covered by both a vertical and horizontal line. Then return to Step 3.

This rather complicated step can be explained as follows. It is equivalent to, first, subtracting this smallest entry from all entries of each row not yet completely covered by a line, whether or not the entry is zero, and second, adding this quantity to every column covered by a line. This second step has the effect of restoring to zero those zero values that just became negative. As we have seen, subtracting the same quantity from every entry of a row does not change the optimal solution; we are just raising the fixed cost charged by certain of the participants. Similarly, adding the same quantity to each entry of a column just increases the cost of the job, regardless of who performs it, so does not change the optimal solution.

Our matrix becomes

$$\begin{bmatrix} 0 & 3 & \mathbf{0} \\ \mathbf{0} & \mathbf{0} & 10 \\ 8 & 0 & \mathbf{0} \end{bmatrix}.$$

In our example, when we return to Step 3 we find that we need three lines now and so we are finished. There are two optimal allocations: one is to assign the first job to the first person, the second job to the second person, and the third job to the third person, for a total cost of 176 dollars; the other optimal allocation is to assign the second person to the first job, the third person to the second job, and the first person to the third job, again with a total cost of 176 dollars.

1.11 Exercises

Ex. 1.1 [177] *Suppose that, in order to reduce automobile gasoline consumption, the government sets a fuel-efficiency target of T km/liter, and then decrees that, if an auto maker produces a make of car with fuel efficiency of $b < T$, then it must also produce a make of car with fuel efficiency rT, for some $r > 1$, such that the average of rT and b is T. Assume that the car maker sells the same number of each make of car. The question is: Is this a good plan? Why or why not? Be specific and quantitative in your answer. Hint: The correct answer is No!.*

Ex. 1.2 *Let A be the arithmetic mean of a finite set of positive numbers, with x the smallest of these numbers, and y the largest. Show that*

$$xy \leq A(x + y - A),$$

with equality if and only if $x = y = A$.

Ex. 1.3 *Some texts call a function $f(x)$* convex *if*

$$f(\alpha x + (1 - \alpha)y) \leq \alpha f(x) + (1 - \alpha)f(y)$$

for all x and y in the domain of the function and for all α in the interval $[0, 1]$. For this exercise, let us call this two-convex. *Show that this definition is equivalent to the one given in Definition 1.1. Hints: First, give the appropriate definition of* three-convex. *Then show that three-convex is equivalent to two-convex; it will help to write*

$$\alpha_1 x_1 + \alpha_2 x_2 = (1 - \alpha_3)\left(\frac{\alpha_1}{(1 - \alpha_3)} x_1 + \frac{\alpha_2}{(1 - \alpha_3)} x_2\right).$$

Finally, use induction on the number N.

Ex. 1.4 *Minimize the function*

$$f(x) = x^2 + \frac{1}{x^2} + 4x + \frac{4}{x},$$

over positive x. *Note that the minimum value of* $f(x,y)$ *cannot be found by a straight-forward application of the AGM Inequality to the four terms taken together. Try to find a way of rewriting* $f(x)$, *perhaps using more than four terms, so that the AGM Inequality can be applied to all the terms.*

Ex. 1.5 *Find the maximum value of* $f(x,y) = x^2y$, *if* x *and* y *are restricted to positive real numbers for which* $6x + 5y = 45$.

Ex. 1.6 *Find the smallest value of*

$$f(x) = 5x + \frac{16}{x} + 21,$$

over positive x.

Ex. 1.7 *Find the smallest value of the function*

$$f(x,y) = \sqrt{x^2 + y^2},$$

among those values of x *and* y *satisfying* $3x - y = 20$.

Ex. 1.8 *Find the maximum and minimum values of the function*

$$f(x) = \sqrt{100 + x^2} - x$$

over nonnegative x.

Ex. 1.9 *Multiply out the product*

$$(x + y + z)\Big(\frac{1}{x} + \frac{1}{y} + \frac{1}{z}\Big)$$

and deduce that the least value of this product, over nonnegative x, y, *and* z, *is* 9. *Use this to find the least value of the function*

$$f(x,y,z) = \frac{1}{x} + \frac{1}{y} + \frac{1}{z},$$

over nonnegative x, y, *and* z *having a constant sum* c.

Ex. 1.10 *The* harmonic mean *of positive numbers* $a_1, ..., a_N$ *is*

$$H = \left(\Big(\frac{1}{a_1} + ... + \frac{1}{a_N}\Big)/N\right)^{-1}.$$

Prove that the geometric mean G *is not less than* H.

Ex. 1.11 *Prove that*

$$\Big(\frac{1}{a_1}+...+\frac{1}{a_N}\Big)(a_1+...+a_N)\geq N^2,$$

with equality if and only if $a_1=...=a_N$.

Ex. 1.12 *Show that the Equation (1.5),* $S=ULU^T$, *can be written as*

$$S=\lambda_1 u^1(u^1)^T+\lambda_2 u^2(u^2)^T+...+\lambda_N u^N(u^N)^T,$$

and

$$S^{-1}=\frac{1}{\lambda_1}u^1(u^1)^T+\frac{1}{\lambda_2}u^2(u^2)^T+...+\frac{1}{\lambda_N}u^N(u^N)^T.$$

Ex. 1.13 *Show that a real symmetric matrix* Q *is nonnegative (positive) definite if and only if all its eigenvalues are nonnegative (positive).*

Ex. 1.14 *Let* Q *be positive-definite, with positive eigenvalues*

$$\lambda_1\geq...\geq\lambda_N>0$$

and associated mutually orthogonal norm-one eigenvectors u^n. *Show that*

$$x^TQx\leq\lambda_1,$$

for all vectors x *with* $\|x\|_2=1$, *with equality if* $x=u^1$. *Hints: Use*

$$1=\|x\|_2^2=x^Tx=x^TIx,$$

$$I=u^1(u^1)^T+...+u^N(u^N)^T,$$

and Equation (1.12).

Ex. 1.15 Young's Inequality *Suppose that* p *and* q *are positive numbers greater than one such that* $\frac{1}{p}+\frac{1}{q}=1$. *If* x *and* y *are positive numbers, then*

$$xy\leq\frac{x^p}{p}+\frac{y^q}{q},$$

with equality if and only if $x^p=y^q$. *Hint: Use the GAGM Inequality.*

Ex. 1.16 [168] *For given constants* c *and* d, *find the largest and smallest values of* $cx+dy$ *taken over all points* (x,y) *of the ellipse*

$$\frac{x^2}{a^2}+\frac{y^2}{b^2}=1.$$

Ex. 1.17 [168] *Find the largest and smallest values of* $2x+y$ *on the circle* $x^2+y^2=1$. *Where do these values occur?*

Ex. 1.18 *When a real* M *by* N *matrix* A *is stored in the computer it is usually* vectorized; *that is, the matrix*

$$A = \begin{bmatrix} A_{11} & A_{12} & \dots & A_{1N} \\ A_{21} & A_{22} & \dots & A_{2N} \\ \cdot & & & \\ \cdot & & & \\ \cdot & & & \\ A_{M1} & A_{M2} & \dots & A_{MN} \end{bmatrix}$$

becomes

$$\mathbf{vec}(A) = (A_{11}, A_{21}, ..., A_{M1}, A_{12}, A_{22}, ..., A_{M2}, ..., A_{MN})^T.$$

Show that the dot product $\mathbf{vec}(A) \cdot \mathbf{vec}(B) = \mathbf{vec}(B)^T \mathbf{vec}(A)$ *can be obtained by*

$$\mathbf{vec}(A) \cdot \mathbf{vec}(B) = \text{trace}\,(AB^T) = \text{trace}\,(B^T A).$$

Ex. 1.19 *Apply the Hungarian Method to solve the allocation problem with the cost matrix*

$$C = \begin{bmatrix} 90 & 75 & 75 & 80 \\ 35 & 85 & 55 & 65 \\ 125 & 95 & 90 & 105 \\ 45 & 110 & 95 & 115 \end{bmatrix}.$$

You should find that the minimum cost is 275 *dollars.*

Chapter 2

Geometric Programming

2.1 Chapter Summary

Geometric Programming (GP) involves the minimization of functions of a special type, known as *posynomials*. The first systematic treatment of geometric programming appeared in the book [101] by Duffin, Peterson and Zener, the founders of geometric programming. As we shall see, the Generalized Arithmetic-Geometric Mean Inequality plays an important role in the theoretical treatment of geometric programming. In this chapter we introduce the notions of duality and cross-entropy distance, and begin our study of iterative algorithms. Some of this discussion of the GP problem follows that in Peressini *et al.* [176].

2.2 An Example of a GP Problem

The following optimization problem was presented originally by Duffin, *et al.* [101] and discussed by Peressini *et al.* in [176]. It illustrates well the type of problem considered in geometric programming. Suppose that 400 cubic yards of gravel must be ferried across a river in an open box of length t_1, width t_2 and height t_3. Each round-trip costs ten cents. The sides and the bottom of the box cost 10 dollars per square yard to build, while the ends of the box cost 20 dollars per square yard. The box will have no salvage value after it has been used. Determine the dimensions of the box that minimize the total cost.

Although we know that the number of trips across the river must be a positive integer, we shall ignore that limitation in what follows, and use $400/t_1t_2t_3$ as the number of trips. In this particular example, it will turn out that this quantity is a positive integer.

With $t = (t_1, t_2, t_3)$, the cost function is

$$g(t) = \frac{40}{t_1t_2t_3} + 20t_1t_3 + 10t_1t_2 + 40t_2t_3, \tag{2.1}$$

which is to be minimized over $t_i > 0$, for $i = 1, 2, 3$. The function $g(t)$ is an example of a posynomial.

2.3 Posynomials and the GP Problem

Functions $g(t)$ of the form

$$g(t) = \sum_{j=1}^{n} c_j \Big(\prod_{i=1}^{m} t_i^{a_{ij}} \Big), \tag{2.2}$$

with $t = (t_1, ..., t_m)$, the $t_i > 0$, $c_j > 0$ and a_{ij} real, are called *posynomials.* The *geometric programming problem*, denoted GP, is to minimize a given posynomial over positive t. In order for the minimum to be greater than zero, we need some of the a_{ij} in Equation (2.2) to be negative.

We denote by $u_j(t)$ the function

$$u_j(t) = c_j \prod_{i=1}^{m} t_i^{a_{ij}},$$

so that

$$g(t) = \sum_{j=1}^{n} u_j(t).$$

For any choice of $\delta_j > 0$, $j = 1, ..., n$, with

$$\sum_{j=1}^{n} \delta_j = 1,$$

we have

$$g(t) = \sum_{j=1}^{n} \delta_j \Big(\frac{u_j(t)}{\delta_j}\Big).$$

Applying the Generalized Arithmetic-Geometric Mean (GAGM) Inequality, we have

$$g(t) \geq \prod_{j=1}^{n} \Big(\frac{u_j(t)}{\delta_j}\Big)^{\delta_j}.$$

Therefore,

$$g(t) \geq \prod_{j=1}^{n} \Big(\frac{c_j}{\delta_j}\Big)^{\delta_j} \Bigg(\prod_{j=1}^{n}\prod_{i=1}^{m} t_i^{a_{ij}\delta_j}\Bigg), \text{ or}$$

$$g(t) \geq \prod_{j=1}^{n} \Big(\frac{c_j}{\delta_j}\Big)^{\delta_j} \Bigg(\prod_{i=1}^{m} t_i^{\sum_{j=1}^{n} a_{ij}\delta_j}\Bigg), \tag{2.3}$$

Suppose that we can find $\delta_j > 0$ with

$$\sum_{j=1}^{n} a_{ij}\delta_j = 0,$$

for each i. We let δ be the vector $\delta = (\delta_1, ..., \delta_n)$. Then the inequality in (2.3) becomes

$$g(t) \geq v(\delta),$$

for

$$v(\delta) = \prod_{j=1}^{n} \Big(\frac{c_j}{\delta_j}\Big)^{\delta_j}. \tag{2.4}$$

Note that we can also write Equation (2.4) as

$$\log v(\delta) = \sum_{j=1}^{n} \delta_j \log\Big(\frac{c_j}{\delta_j}\Big).$$

2.4 The Dual GP Problem

The *dual geometric programming problem*, denoted DGP, is to maximize the function $v(\delta)$, over all *feasible* $\delta = (\delta_1, ..., \delta_n)$, that is, all positive δ for which

$$\sum_{j=1}^{n} \delta_j = 1, \tag{2.5}$$

and

$$\sum_{j=1}^{n} a_{ij}\delta_j = 0, \tag{2.6}$$

for each $i = 1, ..., m$.

Denote by A the $m+1$ by n matrix with entries $A_{ij} = a_{ij}$, and $A_{m+1,j} = 1$, for $j = 1, ..., n$ and $i = 1, ..., m$; an example of such a matrix A occurs in Equation (2.12). Then we can write Equations (2.5) and (2.6) as

$$A\delta = u = \begin{bmatrix} 0 \\ 0 \\ \cdot \\ \cdot \\ \cdot \\ 0 \\ 1 \end{bmatrix}.$$

Clearly, we have

$$g(t) \geq v(\delta), \tag{2.7}$$

for any positive t and feasible δ. Of course, there may be no feasible δ, in which case DGP is said to be *inconsistent*.

As we have seen, the inequality in (2.7) is based on the GAGM Inequality. We have equality in the GAGM Inequality if and only if the terms in the arithmetic mean are all equal. In this case, this says that there is a constant λ such that

$$\frac{u_j(t)}{\delta_j} = \lambda,$$

for each $j = 1, ..., n$. Using the fact that the δ_j sum to one, it follows that

$$\lambda = \sum_{j=1}^{n} u_j(t) = g(t),$$

and

$$\delta_j = \frac{u_j(t)}{g(t)}, \tag{2.8}$$

for each $j = 1, ..., n$.

As the theorem below asserts, if t^* is positive and minimizes $g(t)$, then δ^*, the associated δ from Equation (2.8), is feasible and solves DGP. Since we have equality in the GAGM Inequality now, we have

$$g(t^*) = v(\delta^*).$$

The main theorem in geometric programming is the following.

Theorem 2.1 *If $t^* > 0$ minimizes $g(t)$, then DGP is consistent. In addition, the choice*

$$\delta_j^* = \frac{u_j(t^*)}{g(t^*)} \tag{2.9}$$

is feasible and solves DGP. Finally,

$$g(t^*) = v(\delta^*);$$

that is, there is no duality gap.

Proof: We have

$$\frac{\partial u_j}{\partial t_i}(t^*) = \frac{a_{ij} u_j(t^*)}{t_i^*},$$

so that

$$t_i^* \frac{\partial u_j}{\partial t_i}(t^*) = a_{ij} u_j(t^*), \tag{2.10}$$

for each $i = 1, ..., m$. Since t^* minimizes $g(t)$, we have

$$0 = \frac{\partial g}{\partial t_i}(t^*) = \sum_{j=1}^{n} \frac{\partial u_j}{\partial t_i}(t^*),$$

so that, from Equation (2.10), we have

$$0 = \sum_{j=1}^{n} a_{ij} u_j(t^*),$$

for each $i = 1, ..., m$. It follows that δ^* is feasible. Since

$$u_j(t^*)/\delta_j^* = g(t^*) = \lambda,$$

for all j, we have equality in the GAGM Inequality, and we know

$$g(t^*) = v(\delta^*).$$

Therefore, δ^* solves DGP. This completes the proof. ∎

In Exercise 2.1 you are asked to show that the function

$$g(t_1, t_2) = \frac{2}{t_1 t_2} + t_1 t_2 + t_1$$

has no minimum over the region $t_1 > 0$, and $t_2 > 0$. As you will discover, the DGP is inconsistent in this case. We can still ask if there is a positive greatest lower bound to the values that g can take on. Without too much difficulty, we can determine that if $t_1 \geq 1$ then $g(t_1, t_2) \geq 3$, while if $t_2 \leq 1$ then $g(t_1, t_2) \geq 4$. Therefore, our hunt for the greatest lower bound is concentrated in the region described by $0 < t_1 < 1$, and $t_2 > 1$. Since there is no minimum, we must consider values of t_2 going to infinity, but such that $t_1 t_2$ does not go to infinity and $t_1 t_2$ does not go to zero; therefore, t_1 must go to zero. Suppose we let $t_2 = \frac{f(t_1)}{t_1}$, for some function $f(t)$ such that $f(0) > 0$. Then, as t_1 goes to zero, $g(t_1, t_2)$ goes to $\frac{2}{f(0)} + f(0)$. The exercise asks you to determine how small this limiting quantity can be.

2.5 Solving the GP Problem

The theorem suggests how we might go about solving GP. First, we try to find a feasible δ^* that maximizes $v(\delta)$. This means we have to find a positive solution to the system of $m + 1$ linear equations in n unknowns, given by

$$\sum_{j=1}^{n} \delta_j = 1,$$

and

$$\sum_{j=1}^{n} a_{ij} \delta_j = 0,$$

for $i = 1, ..., m$, such that $v(\delta)$ is maximized. As we shall see, the *multiplicative algebraic reconstruction technique* (MART) is an iterative procedure that we can use to find such δ. If there is no such vector, then GP has no minimizer. Once the desired δ^* has been found, we set

$$\delta_j^* = \frac{u_j(t^*)}{v(\delta^*)},$$

for each $j = 1, ..., n$, and then solve for the entries of t^*. This last step can be simplified by taking logs; then we have a system of linear equations to solve for the values $\log t_i^*$.

2.6 Solving the DGP Problem

The iterative multiplicative algebraic reconstruction technique MART can be used to maximize the function $v(\delta)$, subject to linear equality constraints, provided that the matrix involved has nonnegative entries. We cannot apply the MART yet, because the matrix A does not satisfy these conditions.

2.6.1 The MART

The Kullback–Leibler, or KL distance [142] between positive numbers a and b is

$$KL(a,b) = a\log\frac{a}{b} + b - a. \tag{2.11}$$

We also define $KL(a,0) = +\infty$ and $KL(0,b) = b$. Extending the definition in Equation (2.11) to nonnegative vectors $a = (a_1, ..., a_J)^T$ and $b = (b_1, ..., b_J)^T$, we have

$$KL(a,b) = \sum_{j=1}^{J} KL(a_j, b_j) = \sum_{j=1}^{J} \Big(a_j \log\frac{a_j}{b_j} + b_j - a_j\Big).$$

The MART is an iterative algorithm for finding a nonnegative solution of the system $Px = y$, for an I by J matrix P with nonnegative entries and vector y with positive entries. We also assume that the column sums of P are positive, that is,

$$s_j = \sum_{i=1}^{I} P_{ij} > 0,$$

for all $j = 1, ..., J$. When discussing the MART, we say that the system $Px = y$ is *consistent* when it has nonnegative solutions. We consider two different versions of the MART.

2.6.2 MART I

Both MART algorithms begin with the selection of a positive starting vector x^0. The iterative step of the first version of MART, which we shall call MART I, is the following. For $k = 0, 1, ...,$ and $i = k(\bmod I) + 1$, let

$$x_j^{k+1} = x_j^k \Big(\frac{y_i}{(Px^k)_i}\Big)^{P_{ij}/m_i},$$

for $j = 1, ..., J$, where the parameter m_i is defined to be

$$m_i = \max\{P_{ij} | j = 1, ..., J\}.$$

As $k \to +\infty$, the MART I sequence $\{x^k\}$ converges, in the consistent case, to the nonnegative solution of $y = Px$ for which the KL distance $KL(x, x^0)$ is minimized.

2.6.3 MART II

The iterative step of the second version of MART, which we shall call MART II, is the following. For $k = 0, 1, ...,$ and $i = k(\text{mod}\, I) + 1$, let

$$x_j^{k+1} = x_j^k \Big(\frac{y_i}{(Px^k)_i} \Big)^{P_{ij}/s_j n_i},$$

for $j = 1, ..., J$, where the parameter n_i is defined to be

$$n_i = \max\{P_{ij} s_j^{-1} | j = 1, ..., J\}.$$

The MART II algorithm converges, in the consistent case, to the nonnegative solution for which the KL distance

$$\sum_{j=1}^{J} s_j KL(x_j, x_j^0)$$

is minimized.

2.6.4 Using the MART to Solve the DGP Problem

The entries on the bottom row of A are all one, as is the bottom entry of the column vector u, since these entries correspond to the equation $\sum_{j=1}^{n} \delta_j = 1$. By adding suitably large positive multiples of this last equation to the other equations in the system, we obtain an equivalent system, $B\delta = r$, for which the new matrix B and the new vector r have only positive entries. Now we can apply the MART I algorithm to the system $B\delta = r$, letting $I = m + 1$, $J = n$, $P = B$, $s_j = \sum_{i=1}^{m+1} B_{ij}$, for $j = 1, ..., n$, $\delta = x$, $x^0 = c$ and $y = r$. In the consistent case, the MART I algorithm will find the nonnegative solution that minimizes $KL(x, x^0)$, so we select $x^0 = c$. Then the MART I algorithm finds the nonnegative δ^* satisfying $B\delta^* = r$, or, equivalently, $A\delta^* = u$, for which the KL distance

$$KL(\delta, c) = \sum_{j=1}^{n} \Big(\delta_j \log \frac{\delta_j}{c_j} + c_j - \delta_j \Big)$$

is minimized. Since we know that

$$\sum_{j=1}^{n} \delta_j = 1,$$

it follows that minimizing $KL(\delta, c)$ is equivalent to maximizing $v(\delta)$. Using δ^*, we find the optimal t^* solving the GP problem.

For example, the linear system of equations $A\delta = u$ corresponding to the posynomial in Equation (2.1) is

$$A\delta = u = \begin{bmatrix} -1 & 1 & 1 & 0 \\ -1 & 0 & 1 & 1 \\ -1 & 1 & 0 & 1 \\ 1 & 1 & 1 & 1 \end{bmatrix} \begin{bmatrix} \delta_1 \\ \delta_2 \\ \delta_3 \\ \delta_4 \end{bmatrix} = \begin{bmatrix} 0 \\ 0 \\ 0 \\ 1 \end{bmatrix}. \tag{2.12}$$

Adding two times the last row to the other rows, the system becomes

$$B\delta = r = \begin{bmatrix} 1 & 3 & 3 & 2 \\ 1 & 2 & 3 & 3 \\ 1 & 3 & 2 & 3 \\ 1 & 1 & 1 & 1 \end{bmatrix} \begin{bmatrix} \delta_1 \\ \delta_2 \\ \delta_3 \\ \delta_4 \end{bmatrix} = \begin{bmatrix} 2 \\ 2 \\ 2 \\ 1 \end{bmatrix}. \tag{2.13}$$

The matrix B and the vector r are now positive. We are ready to apply the MART to the system in Equation (2.13).

The MART iteration is as follows. With $i = k(\bmod (m+1)) + 1$, $m_i = \max\{B_{ij} \,|\, j = 1, 2, ..., n\}$ and $k = 0, 1, ...$, let

$$\delta_j^{k+1} = \delta_j^k \left(\frac{r_i}{(B\delta^k)_i} \right)^{m_i^{-1} B_{ij}}.$$

Using the MART, beginning with $\delta^0 = c$, we find that the optimal δ^* is $\delta^* = (0.4, 0.2, 0.2, 0.2)^T$. Now we find $v(\delta^*)$, which, by Theorem 2.1, equals $g(t^*)$.

We have

$$v(\delta^*) = \left(\frac{40}{0.4}\right)^{0.4} \left(\frac{20}{0.2}\right)^{0.2} \left(\frac{10}{0.2}\right)^{0.2} \left(\frac{40}{0.2}\right)^{0.2},$$

so that, after a little arithmetic, we discover that $v(\delta^*) = g(t^*) = 100$; the lowest cost is one hundred dollars.

Using Equation (2.9) for $i = 1, ..., 4$, we have

$$u_1(t^*) = \frac{40}{t_1^* t_2^* t_3^*} = 100\delta_1^* = 40,$$

$$u_2(t^*) = 20 t_1^* t_3^* = 100\delta_2^* = 20,$$

$$u_3(t^*) = 10 t_1^* t_2^* = 100\delta_3^* = 20,$$

and

$$u_4(t^*) = 40 t_2^* t_3^* = 100\delta_4^* = 20.$$

Again, a little arithmetic reveals that $t_1^* = 2$, $t_2^* = 1$, and $t_3^* = 0.5$. Here we were able to solve the system of nonlinear equations fairly easily. Generally, however, we will need to take logarithms of both sides of each equation, and then solve the resulting system of linear equations for the unknowns $x_i^* = \log t_i^*$.

2.7 Constrained Geometric Programming

Consider now the following variant of the problem of transporting the gravel across the river. Suppose that the bottom and the two sides will be constructed for free from scrap metal, but only four square yards are available. The cost function to be minimized becomes

$$g_0(t) = \frac{40}{t_1 t_2 t_3} + 40 t_2 t_3,$$

and the constraint is

$$g_1(t) = \frac{t_1 t_3}{2} + \frac{t_1 t_2}{4} \leq 1.$$

With $\delta_1 > 0$, $\delta_2 > 0$, and $\delta_1 + \delta_2 = 1$, we write

$$g_0(t) = \delta_1 \frac{40}{\delta_1 t_1 t_2 t_3} + \delta_2 \frac{40 t_2 t_3}{\delta_2}.$$

Since $0 \leq g_1(t) \leq 1$, we have

$$g_0(t) \geq \Big(\delta_1 \frac{40}{\delta_1 t_1 t_2 t_3} + \delta_2 \frac{40 t_2 t_3}{\delta_2}\Big)\Big(g_1(t)\Big)^{\lambda},$$

for any positive λ. The GAGM Inequality then tells us that

$$g_0(t) \geq \Bigg(\Big(\frac{40}{\delta_1 t_1 t_2 t_3}\Big)^{\delta_1}\Big(\frac{40 t_2 t_3}{\delta_2}\Big)^{\delta_2}\Bigg)\Big(g_1(t)\Big)^{\lambda},$$

so that

$$g_0(t) \geq \Bigg(\Big(\frac{40}{\delta_1}\Big)^{\delta_1}\Big(\frac{40}{\delta_2}\Big)^{\delta_2}\Bigg) t_1^{-\delta_1} t_2^{\delta_2 - \delta_1} t_3^{\delta_2 - \delta_1}\Big(g_1(t)\Big)^{\lambda}. \tag{2.14}$$

From the GAGM Inequality, we also know that, for $\delta_3 > 0$, $\delta_4 > 0$ and $\lambda = \delta_3 + \delta_4$,

$$\Big(g_1(t)\Big)^{\lambda} \geq (\lambda)^{\lambda}\Bigg(\Big(\frac{1}{2\delta_3}\Big)^{\delta_3}\Big(\frac{1}{4\delta_4}\Big)^{\delta_4}\Bigg) t_1^{\delta_3+\delta_4} t_2^{\delta_4} t_3^{\delta_3}. \tag{2.15}$$

Combining the inequalities in (2.14) and (2.15), and writing $\delta = (\delta_1, \delta_2, \delta_3, \delta_4)$, we obtain

$$g_0(t) \geq v(\delta) t_1^{-\delta_1+\delta_3+\delta_4} t_2^{-\delta_1+\delta_2+\delta_4} t_3^{-\delta_1+\delta_2+\delta_3},$$

with

$$v(\delta) = \left(\frac{40}{\delta_1}\right)^{\delta_1}\left(\frac{40}{\delta_2}\right)^{\delta_2}\left(\frac{1}{2\delta_3}\right)^{\delta_3}\left(\frac{1}{4\delta_4}\right)^{\delta_4}\left(\delta_3+\delta_4\right)^{\delta_3+\delta_4}.$$

If we can find a positive vector δ with

$$\begin{aligned}\delta_1 + \delta_2 &= 1,\\ -\delta_1 + \delta_3 + \delta_4 &= 0,\\ -\delta_1 + \delta_2 + \delta_4 &= 0\\ -\delta_1 + \delta_2 + \delta_3 &= 0,\end{aligned} \tag{2.16}$$

then

$$g_0(t) \geq v(\delta).$$

In this particular case, there is a unique positive δ satisfying the equations (2.16), namely

$$\delta_1^* = \frac{2}{3}, \delta_2^* = \frac{1}{3}, \delta_3^* = \frac{1}{3}, \text{and } \delta_4^* = \frac{1}{3},$$

and

$$v(\delta^*) = 60.$$

Therefore, $g_0(t)$ is bounded below by 60. If there is t^* such that

$$g_0(t^*) = 60,$$

then we must have

$$g_1(t^*) = 1,$$

and equality in the GAGM Inequality. Consequently,

$$\frac{3}{2}\frac{40}{t_1^* t_2^* t_3^*} = 3(40 t_2^* t_3^*) = 60,$$

and

$$\frac{3}{2}t_1^* t_3^* = \frac{3}{4}t_1^* t_2^* = K.$$

Since $g_1(t^*) = 1$, we must have $K = \frac{3}{2}$. We solve these equations by taking logarithms, to obtain the solution

$$t_1^* = 2, t_2^* = 1, \text{and } t_3^* = \frac{1}{2}.$$

The change of variables $t_i = e^{x_i}$ converts the constrained GP problem into a constrained convex programming problem. The theory of the constrained GP problem can then be obtained as a consequence of the theory for the convex problem, which we shall consider in Chapter 10.

2.8 Exercises

Ex. 2.1 *Show that there is no solution to the problem of minimizing the function*

$$g(t_1, t_2) = \frac{2}{t_1 t_2} + t_1 t_2 + t_1,$$

over $t_1 > 0$, $t_2 > 0$. *Can* $g(t_1, t_2)$ *ever be smaller than* $2\sqrt{2}$*?*

Ex. 2.2 *Minimize the function*

$$g(t_1, t_2) = \frac{1}{t_1 t_2} + t_1 t_2 + t_1 + t_2,$$

over $t_1 > 0$, $t_2 > 0$. *This will require some iterative numerical method for solving equations.*

Ex. 2.3 *Program the MART algorithm and use it to verify the assertions made previously concerning the solutions of the two numerical examples.*

Chapter 3

Basic Analysis

3.1 Chapter Summary

The theory and practice of continuous optimization relies heavily on the basic notions and tools of real analysis. In this chapter we review important topics from analysis that we shall need later.

3.2 Minima and Infima

When we say that we seek the minimum value of a function $f(x)$ over x within some set C we imply that there is a point z in C such that $f(z) \leq f(x)$ for all x in C. Of course, this need not be the case. For example, take the function $f(x) = x$ defined on the real numbers and C the set of positive real numbers. In such cases, instead of looking for the minimum of $f(x)$ over x in C, we may seek the *infimum* or *greatest lower bound* of the values $f(x)$, over x in C.

Definition 3.1 *We say that a number α is the* infimum *of a subset S of $\mathbb{R}$, abbreviated $\alpha = \inf(S)$, or the* greatest lower bound *of S, abbreviated $\alpha = \text{glb}(S)$, if two conditions hold:*

(1) $\alpha \leq s$, *for all* s *in* S; *and*

(2) if $t \leq s$ *for all* s *in* S, *then* $t \leq \alpha$.

Definition 3.2 *We say that a number* β *is the* supremum *of a subset* S *in* $\mathbb{R}$, *abbreviated* $\beta = \sup(S)$, *or the* least upper bound *of* S, *abbreviated* $\beta = \text{lub}(S)$, *if two conditions hold:*

(1) $\beta \geq s$, *for all* s *in* S; *and*

(2) if $t \geq s$ *for all* s *in* S, *then* $t \geq \beta$.

In our example of $f(x) = x$ and C as the set of positive real numbers, let $S = \{f(x)|x \in C\}$. Then the infimum of S is $\alpha = 0$, although there is no s in S for which $s = 0$. Whenever there is a point z in C with $\alpha = f(z)$, then $f(z)$ is both the infimum and the minimum of $f(x)$ over x in C.

3.3 Limits

We begin with the basic definitions pertaining to limits. Concerning notation, we denote by x a member of $\mathbb{R}^J$, so that, for $J = 1$, x will denote a real number. Members x of $\mathbb{R}^J$ will always be thought of as column vectors, so that x^T, the transpose of x, is a row vector. Entries of an x in $\mathbb{R}^J$ we denote by x_j, so x_j will always denote a real number; in contrast, x^k will denote a member of $\mathbb{R}^J$, with entries x_j^k.

For a vector x in $\mathbb{R}^J$ we shall denote by $\|x\|$ an arbitrary norm. The notation $\|x\|_2$ will always refer to the two-norm of a vector x; that is,

$$\|x\|_2 = \sqrt{\sum_{j=1}^{J} |x_j|^2}.$$

The two-norm of x is the Euclidean distance from the point x to the origin, or, equivalently, the length of the directed line segment from the origin to x. Associated with the two-norm is the inner product

$$\langle x, y \rangle = \sum_{j=1}^{J} x_j y_j.$$

We sometimes write the inner product as a dot product or using matrix multiplication:

$$\langle x, y \rangle = x \cdot y = x^T y.$$

Note that

$$\|x\|_2^2 = \langle x, x \rangle.$$

The two-norm is not the only interesting norm on $\mathbb{R}^J$, though. Another one is the one-norm,

$$\|x\|_1 = \sum_{j=1}^{J} |x_j|.$$

Any norm is a generalization of the notion of absolute value of a real number; for any real number x we can view $|x|$ as the distance from x to 0. For real numbers x and z, $|x - z|$ is the distance from x to z. For points x and z in $\mathbb{R}^J$, $\|x - z\|$ should be viewed as the distance from the point x to the point z, or, equivalently, the length of the directed line segment from z to x; each norm defines a different notion of distance.

In the definitions that follow we use an arbitrary norm on $\mathbb{R}^J$. The reason for this is that these definitions are independent of the particular norm used. A sequence is bounded, Cauchy, or convergent with respect to one norm if and only if it is the same with respect to any norm. Similarly, a function is continuous with respect to one norm if and only if it is continuous with respect to any other norm.

Definition 3.3 *A sequence $\{x^n | n = 1, 2, ...\}$, $x^n \in \mathbb{R}^J$, is said to* converge *to $z \in \mathbb{R}^J$, or have* limit *z if, given any $\epsilon > 0$, there is $N = N(\epsilon)$, usually depending on ϵ, such that*

$$\|x^n - z\| \leq \epsilon,$$

whenever $n \geq N(\epsilon)$.

Definition 3.4 *A sequence $\{x^n\}$ in $\mathbb{R}^J$ is* bounded *if there is a constant B such that $\|x^n\| \leq B$, for all n.*

It is convenient to extend the notion of limit of a sequence of real numbers to include the infinities.

Definition 3.5 *A sequence of real numbers $\{x_n | n = 1, 2, ...\}$ is said to* converge to $+\infty$ *if, given any $b > 0$, there is $N = N(b)$, usually depending on b, such that $x_n \geq b$, whenever $n \geq N(b)$. A sequence of real numbers $\{x_n | n = 1, 2, ...\}$ is said to* converge to $-\infty$ *if the sequence $\{-x_n\}$ converges to $+\infty$.*

Definition 3.6 *Let $f : \mathbb{R}^J \to \mathbb{R}^M$. We say that $z \in \mathbb{R}^M$ is the* limit *of $f(x)$, as $x \to a$ in $\mathbb{R}^J$, if, for every sequence $\{x^n\}$ converging to a, with $x^n \neq a$ for all n, the sequence $\{f(x^n)\}$ in $\mathbb{R}^M$ converges to z. We then write*

$$z = \lim_{x \to a} f(x).$$

For $M = 1$, we allow z to be infinite.

Definition 3.7 *A subset C of $\mathbb{R}^J$ is* bounded *if there is a positive number B such that $\|x\| \leq B$, for all x in C. A subset C is* closed *if, whenever there is a sequence $\{x^n|n = 1, 2, ...\}$, with each x^n in C and the sequence converging to x, the vector x is also in C.*

3.4 Completeness

One version of the axiom of completeness for the set of real numbers $\mathbb{R}$ is that every nonempty subset of $\mathbb{R}$ that is bounded above has a least upper bound, or, equivalently, every nonempty subset of $\mathbb{R}$ that is bounded below has a greatest lower bound. The notion of completeness is usually not emphasized in beginning calculus courses and encountered for the first time in a real analysis course. But without completeness, many of the fundamental theorems in calculus would not hold. If we tried to do calculus by considering only rational numbers, the intermediate value theorem would not hold, and it would be possible for a differentiable function to have a positive derivative without being increasing.

To further illustrate the importance of completeness, consider the proof of the following proposition.

Proposition 3.1 *The sequence $\{\frac{1}{n}\}$ converges to zero, as $n \rightarrow +\infty$.*

Suppose we attempt to prove this proposition simply by applying the definition of the limit of a sequence. Let $\epsilon > 0$ be given. Select a positive integer N with $N > \frac{1}{\epsilon}$. Then, whenever $n \geq N$, we have

$$|\frac{1}{n} - 0| = \frac{1}{n} \leq \frac{1}{N} < \epsilon.$$

This would seem to complete the proof of the proposition. But it is incorrect. The flaw in the argument is in the choice of N. We do not yet know that we can select N with $N > \frac{1}{\epsilon}$, since this is equivalent to $\frac{1}{N} < \epsilon$. Until we know that the proposition is true, we do not know that we can make $\frac{1}{N}$ as small as desired by the choice of N. The proof requires completeness.

Let S be the set $\{1, \frac{1}{2}, \frac{1}{3}, \frac{1}{4}, ...\}$. This set is nonempty and bounded below by any negative real number. Therefore, by completeness, S has a greatest lower bound; call it L. It is not difficult to prove that the decreasing sequence $\{\frac{1}{n}\}$ must then converge to L, and the subsequence $\{\frac{1}{2n}\}$ must also converge to L. But since the limit of a product is the product of the limits, whenever all the limits exist, we also know that the sequence $\{\frac{1}{2n}\}$ converges to $\frac{L}{2}$. Therefore, $L = \frac{L}{2}$, and $L = 0$ must follow. Now the proof is complete.

The rational number line has "holes" in it that the irrational numbers fill; in this sense, the completeness of the real numbers is sometimes characterized by saying that it has no holes in it. But the completeness of the reals actually tells us other things about the structure of the real numbers. We know, for example, that there are no rational numbers that are larger than all the positive integers. But can there be irrational numbers that are larger than all the positive integers? Completeness tells us that the answer is no.

Corollary 3.1 *There is no real number larger than all the positive integers.*

Proof: Suppose, to the contrary, that there is a real number b such that $b > n$, for all positive integers n. Then $0 < \frac{1}{b} < \frac{1}{n}$, for all positive integers n. But this cannot happen, since, by the previous proposition, $\{\frac{1}{n}\}$ converges to zero. ∎

Notice that, if we restrict ourselves to the world of rational numbers when we define the concept of limit of a sequence, then we must also restrict the ϵ to the rationals; suppose we call this the "rational limit."When we do this, we can show that the sequence $\{\frac{1}{n}\}$ converges to zero. What we have really shown with the proposition and corollary above is that, if a sequence of rational numbers converges to a rational number, in the sense of the "rational limit,"then it converges to that rational number in the usual sense as well.

For the more general spaces $\mathbb{R}^J$ completeness is expressed, for example, by postulating that every Cauchy sequence is a convergent sequence.

Definition 3.8 *A sequence $\{x^n\}$ of vectors in $\mathbb{R}^J$ is called a* Cauchy sequence *if, for every $\epsilon > 0$ there is a positive integer $N = N(\epsilon)$, usually depending on ϵ, such that, for all m and n greater than N, we have $\|x^n - x^m\| < \epsilon$.*

Every convergent sequence in $\mathbb{R}^J$ is bounded and is a Cauchy sequence. The Bolzano–Weierstrass Theorem tells us that every bounded sequence in $\mathbb{R}^J$ has a convergent subsequence; this is equivalent to the completeness of the metric space $\mathbb{R}^J$.

Theorem 3.1 (The Bolzano–Weierstrass Theorem) *Let $\{x^n\}$ be a bounded sequence of vectors in $\mathbb{R}^J$. Then $\{x^n\}$ has a convergent subsequence.*

As we shall see, the Bolzano–Weierstrass Theorem plays an important role in proving convergence of iterative methods.

3.5 Continuity

A basic notion in analysis is that of a continuous function. Although we shall be concerned primarily with functions whose values are real numbers, we can define continuity for functions whose values lie in $\mathbb{R}^M$.

Definition 3.9 *We say the function $f : \mathbb{R}^J \to \mathbb{R}^M$ is* continuous *at $x = a$ if*

$$f(a) = \lim_{x \to a} f(x).$$

A basic theorem in real analysis is the following:

Theorem 3.2 *Let $f : \mathbb{R}^J \to \mathbb{R}$ be continuous and let C be nonempty, closed, and bounded. Then there are a and b in C with $f(a) \leq f(x)$ and $f(b) \geq f(x)$, for all x in C.*

We give some examples:

(1) The function $f(x) = x$ is continuous and the set $C = [0, 1]$ is nonempty, closed and bounded. The minimum occurs at $x = 0$ and the maximum occurs at $x = 1$.

(2) The set $C = (0, 1]$ is not closed. The function $f(x) = x$ has no minimum value on C, but does have a maximum value $f(1) = 1$.

(3) The set $C = (-\infty, 0]$ is not bounded and $f(x) = x$ has no minimum value on C. Note also that $f(x) = x$ has no finite infimum with respect to C.

Definition 3.10 *Let $f : D \subseteq \mathbb{R}^J \to \mathbb{R}$. For any real α, the* level set *of f corresponding to α is the set $\{x | f(x) \leq \alpha\}$.*

Proposition 3.2 (Weierstrass) *Suppose that $f : D \subseteq \mathbb{R}^J \to R$ is continuous, where D is nonempty and closed, and that every level set of f is bounded. Then f has a global minimizer.*

Proof: This is a standard application of the Bolzano–Weierstrass Theorem. ∎

3.6 Limsup and Liminf

Some of the functions we shall be interested in may be discontinuous at some points. For that reason, it is common in optimization to consider *semi-continuity*, which is weaker than continuity. While continuity involves limits, semi-continuity involves superior and inferior limits.

We know that a real-valued function $f(x) : \mathbb{R}^J \to \mathbb{R}$ is continuous at $x = a$ if, given any $\epsilon > 0$, there is a $\delta > 0$ such that $\|x - a\| < \delta$ implies that $|f(x) - f(a)| < \epsilon$. We then write

$$f(a) = \lim_{x \to a} f(x).$$

We can generalize this notion as follows.

Definition 3.11 *We say that a finite real number β is the* superior limit *or* lim sup *of $f(x)$, as x approaches a, written $\beta = \limsup_{x \to a} f(x)$ if,*

(1) for every $\epsilon > 0$, there is $\delta > 0$ such that, for every x satisfying $\|x - a\| < \delta$, we have $f(x) < \beta + \epsilon$, and

(2) for every $\epsilon > 0$ and $\delta > 0$ there is x with $\|x - a\| < \delta$ and $f(x) > \beta - \epsilon$.

Definition 3.12 *We say that a finite real number α is the* inferior limit *or* lim inf *of $f(x)$, as x approaches a, written $\alpha = \liminf_{x \to a} f(x)$ if,*

(1) for every $\epsilon > 0$, there is $\delta > 0$ such that, for every x satisfying $\|x - a\| < \delta$, we have $f(x) > \alpha - \epsilon$, and

(2) for every $\epsilon > 0$ and $\delta > 0$ there is x with $\|x - a\| < \delta$ and $f(x) < \alpha + \epsilon$.

We leave it as Exercise 3.4 for the reader to show that $\alpha = \liminf_{x \to a} f(x)$ is the largest real number γ with the following property: for every $\epsilon > 0$, there is $\delta > 0$ such that, if $\|x - a\| < \delta$, then $f(x) > \gamma - \epsilon$.

Definition 3.13 *We say that $\beta = +\infty$ is the* superior limit *or* lim sup *of $f(x)$, as x approaches a, written $+\infty = \limsup_{x \to a} f(x)$ if, for every $B > 0$ and $\delta > 0$ there is x with $\|x - a\| < \delta$ and $f(x) > B$.*

Definition 3.14 *We say that $\alpha = -\infty$ is the* inferior limit *or* lim inf *of $f(x)$, as x approaches a, written $-\infty = \liminf_{x \to a} f(x)$ if, for every $B > 0$ and $\delta > 0$, there is x with $\|x - a\| < \delta$ and $f(x) < -B$.*

It follows from the definitions that $\alpha \leq f(a) \leq \beta$.

For example, suppose that $a = 0$, $f(x) = 0$, for $x \neq 0$, and $f(0) = 1$. Then $\beta = 1$ and $\alpha = 0$. If $a = 0$, $f(x) = -1/x$ for $x < 0$ and $f(x) = 1/x$ for $x > 0$, then $\alpha = -\infty$ and $\beta = +\infty$.

It is not immediately obvious that β and α always exist. The next section provides another view of these notions, from which it becomes clear that the existence of β and α is a consequence of the completeness of the space $\mathbb{R}$.

3.7 Another View

We can define the superior and inferior limits in terms of sequences. We leave it to the reader to show that these definitions are equivalent to the ones just given.

Let $f : \mathbb{R}^J \to \mathbb{R}$ and a be fixed in $\mathbb{R}^J$. Let L be the set consisting of all γ, possibly including the infinities, having the property that there is a sequence $\{x^n\}$ in $\mathbb{R}^J$ converging to a such that $\{f(x^n)\}$ converges to γ. It is convenient, now, to permit the sequence $x^n = a$ for all n, so that $\gamma = f(a)$ is in L and L is never empty. Therefore, we always have

$$-\infty \leq \inf(L) \leq f(a) \leq \sup(L) \leq +\infty.$$

For example, let $f(x) = 1/x$ for $x \neq 0$, $f(0) = 0$, and $a = 0$. Then $L = \{-\infty, 0, +\infty\}$, $\inf(L) = -\infty$, and $\sup(L) = +\infty$.

Definition 3.15 *The (possibly infinite) number* $\inf(L)$ *is called the* inferior limit *or* lim inf *of* $f(x)$*, as* $x \to a$ *in* $\mathbb{R}^J$*. The (possibly infinite) number* $\sup(L)$ *is called the* superior limit *or* lim sup *of* $f(x)$*, as* $x \to a$ *in* $\mathbb{R}^J$*.*

It follows from these definitions and our previous discussion that

$$\liminf_{x \to a} f(x) \leq f(a) \leq \limsup_{x \to a} f(x).$$

For example, let $f(x) = x$ for $x < 0$ and $f(x) = x + 1$ for $x > 0$. Then we have

$$\limsup_{x \to 0} f(x) = 1,$$

and

$$\liminf_{x \to 0} f(x) = 0.$$

Proposition 3.3 *The inferior limit and the superior limit are in the set* L*.*

Proof: We leave the proof as Exercise 3.6. ∎

The function doesn't have to be defined at a point in order for the lim sup and lim inf to be defined there. If $f : (0, \delta) \to \mathbb{R}$, for some $\delta > 0$, we have the following definitions:

$$\limsup_{t \downarrow 0} f(t) = \lim_{t \downarrow 0} \Big(\sup\{f(x) | 0 < x < t\} \Big),$$

and

$$\liminf_{t \downarrow 0} f(t) = \lim_{t \downarrow 0} \Big(\inf\{f(x) | 0 < x < t\} \Big).$$

3.8 Semi-Continuity

We know that $\alpha \leq f(a) \leq \beta$. We can generalize the notion of continuity by replacing the limit with the inferior or superior limit. When $M = 1$, $f(x)$ is continuous at $x = a$ if and only if

$$\liminf_{x \to a} f(x) = \limsup_{x \to a} f(x) = f(a).$$

Definition 3.16 *We say that $f : \mathbb{R}^J \to \mathbb{R}$ is* lower semi-continuous *(LSC) at $x = a$ if*

$$f(a) = \alpha = \liminf_{x \to a} f(x).$$

Definition 3.17 *We say that $f : \mathbb{R}^J \to \mathbb{R}$ is* upper semi-continuous *(USC) at $x = a$ if*

$$f(a) = \beta = \limsup_{x \to a} f(x).$$

Note that, if $f(x)$ is LSC (USC) at $x = a$, then $f(x)$ remains LSC (USC) when $f(a)$ is replaced by any lower (higher) value. See Exercise 3.3 for an equivalent definition of lower semi-continuity.

The following theorem of Weierstrass extends Theorem 3.2 and shows the importance of lower semi-continuity for minimization problems.

Theorem 3.3 *Let $f : \mathbb{R}^J \to \mathbb{R}$ be LSC and let C be nonempty, closed, and bounded. Then there is a in C with $f(a) \leq f(x)$, for all x in C.*

3.9 Exercises

Ex. 3.1 *Let S and T be nonempty subsets of the real line, with $s \leq t$ for every s in S and t in T. Prove that* $\mathrm{lub}(S) \leq \mathrm{glb}(T)$.

Ex. 3.2 *Let $f(x, y) : \mathbb{R}^2 \to \mathbb{R}$, and, for each fixed y, let $\inf_x f(x, y)$ denote the greatest lower bound of the set of numbers $\{f(x, y) | x \in \mathbb{R}\}$. Show that*

$$\inf_x \left(\inf_y f(x, y) \right) = \inf_y \left(\inf_x f(x, y) \right).$$

Hint: Note that

$$\inf_y f(x, y) \leq f(x, y),$$

for all x and y.

Ex. 3.3 *Prove that $f : \mathbb{R}^J \to \mathbb{R}$ is lower semi-continuous at $x = a$ if and only if, for every $\epsilon > 0$, there is $\delta > 0$ such that $\|x - a\| < \delta$ implies that $f(x) > f(a) - \epsilon$.*

Ex. 3.4 *Show that $\gamma = \alpha = \liminf_{x \to a} f(x)$ is the largest real number γ with the following property: for every $\epsilon > 0$, there is $\delta > 0$ such that, if $\|x - a\| < \delta$, then $f(x) > \gamma - \epsilon$.*

Ex. 3.5 *Consider the function $f(x)$ defined by $f(x) = e^{-x}$, for $x > 0$ and by $f(x) = -e^x$, for $x < 0$. Show that*

$$-1 = \liminf_{x \to 0} f(x)$$

and

$$1 = \limsup_{x \to 0} f(x).$$

Ex. 3.6 *For $n = 1, 2, ...,$ let*

$$A_n = \left\{ x | \ \|x - a\| \le \frac{1}{n} \right\},$$

and let α_n and β_n be defined by

$$\alpha_n = \inf \left\{ f(x) | \, x \in A_n \right\},$$

and

$$\beta_n = \sup \left\{ f(x) | \, x \in A_n \right\}.$$

(a) Show that the sequence $\{\alpha_n\}$ is increasing, bounded above by $f(a)$ and converges to some α, while the sequence $\{\beta_n\}$ is decreasing, bounded below by $f(a)$ and converges to some β. Hint: Use the fact that, if $A \subseteq B$, where A and B are sets of real numbers, then $\inf(A) \ge \inf(B)$.

(b) Show that α and β are in L. Hint: Prove that there is a sequence $\{x^n\}$ with x^n in A_n and $f(x^n) \le \alpha_n + \frac{1}{n}$.

(c) Show that, if $\{x^m\}$ is any sequence converging to a, then there is a subsequence, denoted $\{x^{m_n}\}$, such that x^{m_n} is in A_n, for each n.

(d) Show that, if $\{f(x^m)\}$ converges to γ, then

$$\alpha_n \le f(x^{m_n}) \le \beta_n,$$

so that

$$\alpha \le \gamma \le \beta.$$

(e) Show that

$$\alpha = \liminf_{x \to a} f(x)$$

and

$$\beta = \limsup_{x \to a} f(x).$$

Chapter 4

Convex Sets

4.1 Chapter Summary

Convex sets and convex functions play important roles in optimization. In this chapter we survey the basic facts concerning convex sets, beginning with the geometry and topology of $\mathbb{R}^J$. We then define convex functions in terms of convex sets. We close with several Theorems of the Alternative concerning linear inequalities.

4.2 The Geometry of Real Euclidean Space

We denote by $\mathbb{R}^J$ the real Euclidean space consisting of all J-dimensional column vectors $x = (x_1, ..., x_J)^T$ with real entries x_j; here the superscript T denotes the transpose of the 1 by J matrix (or, row vector) $(x_1, ..., x_J)$.

4.2.1 Inner Products

For $x = (x_1, ..., x_J)^T$ and $y = (y_1, ..., y_J)^T$ in $\mathbb{R}^J$, the dot product $x \cdot y$ is defined to be

$$x \cdot y = \sum_{j=1}^{J} x_j y_j.$$

Note that we can write

$$x \cdot y = y^T x = x^T y,$$

where juxtaposition indicates matrix multiplication. The 2-norm, or *Euclidean norm*, or *Euclidean length*, of x is

$$||x||_2 = \sqrt{x \cdot x} = \sqrt{x^T x}.$$

The *Euclidean distance* between two vectors x and y in $\mathbb{R}^J$ is $||x - y||_2$.

The space $\mathbb{R}^J$, along with its dot product, is an example of a finite-dimensional Hilbert space.

Definition 4.1 *Let $\mathcal{V}$ be a real vector space. The scalar-valued function $\langle u, v\rangle$ is called an* inner product *on $\mathcal{V}$ if the following four properties hold, for all u, w, and v in $\mathcal{V}$, and all real c:*

$$\begin{aligned} \langle u + w, v\rangle &= \langle u, v\rangle + \langle w, v\rangle; \\ \langle cu, v\rangle &= c\langle u, v\rangle; \\ \langle v, u\rangle &= \langle u, v\rangle; \text{ and} \\ \langle u, u\rangle &\geq 0, \end{aligned} \tag{4.1}$$

with equality in (4.1) if and only if $u = 0$.

The dot product of vectors is an example of an inner product. The properties of an inner product are precisely the ones needed to prove Cauchy's Inequality, which then holds for any inner product. We shall favor the dot product notation $u \cdot v$ for the inner product of vectors, although we shall occasionally use the matrix multiplication form, $v^T u$ or the inner product notation $\langle u, v\rangle$.

4.2.2 Cauchy's Inequality

Cauchy's Inequality, also called the Cauchy–Schwarz Inequality, tells us that

$$|\langle x, y\rangle| \leq ||x||_2||y||_2,$$

with equality if and only if $y = \alpha x$, for some scalar α. The Cauchy–Schwarz Inequality holds for any inner product. We say that the vectors x and y are *mutually orthogonal* if $\langle x, y\rangle = 0$. An alternative approach to orthogonality is presented in Exercise 4.1.

A simple application of Cauchy's inequality gives us

$$||x + y||_2 \leq ||x||_2 + ||y||_2, \tag{4.2}$$

with equality if and only if one of the vectors is a nonnegative multiple of the other one; the inequality in (4.2) is called the *Triangle Inequality.*

The *Parallelogram Law* is an easy consequence of the definition of the 2-norm:

$$||x + y||_2^2 + ||x - y||_2^2 = 2||x||_2^2 + 2||y||_2^2. \tag{4.3}$$

It is important to remember that Cauchy's Inequality and the Parallelogram Law in Equation (4.3) hold only for the 2-norm. One consequence of the Parallelogram Law that we shall need later is the following: if $x \neq y$ and $\|x\|_2 = \|y\|_2 = d$, then $\|\frac{1}{2}(x + y)\|_2 < d$. (Draw a picture!)

4.2.3 Other Norms

The two-norm is the most important, but not the only, norm on the space $\mathbb{R}^J$ that we study; we shall also be interested in the one-norm (see Exercise 4.5). The purely topological results we discuss in the next section are independent of the choice of norm on $\mathbb{R}^J$, and we shall remind the reader of this by using the notation $\|x\|$ to denote an arbitrary norm. Theorems concerning orthogonal projection hold only for the two-norm, which we shall denote by $\|x\|_2$. In fact, whenever we use the word "orthogonal," we shall imply that we are speaking about the two-norm. There have been attempts to define orthogonality in the absence of an inner product, and so for other norms, but the theory here is not as successful.

4.3 A Bit of Topology

Having a norm allows us to define the distance between two points x and y in $\mathbb{R}^J$ as $||x - y||$. Being able to talk about how close points are to each other enables us to define continuity of functions on $\mathbb{R}^J$ and to consider topological notions of closed set, open set, interior of a set and boundary of a set. None of these notions depend on the particular norm we are using.

Definition 4.2 *A subset B of $\mathbb{R}^J$ is* closed *if, whenever x^k is in B for each nonnegative integer k and $||x - x^k|| \to 0$, as $k \to +\infty$, then x is in B.*

For example, $B = [0, 1]$ is closed as a subset of R, but $B = (0, 1)$ is not.

Definition 4.3 *We say that $d \geq 0$ is the* distance from the point x to the set B *if, for every $\epsilon > 0$, there is b_ϵ in B, with $||x - b_\epsilon|| < d + \epsilon$, and no b in B with $||x - b|| < d$.*

The Euclidean distance from the point 0 in R to the set $(0, 1)$ is zero, while its distance to the set $(1, 2)$ is one. It follows easily from the definitions that, if B is closed and $d = 0$, then x is in B.

Definition 4.4 *The* closure *of a set B is the set of all points x whose distance from B is zero.*

The closure of the interval $B = (0, 1)$ is $[0, 1]$.

Definition 4.5 *A subset U of $\mathbb{R}^J$ is* open *if its complement, the set of all points not in U, is closed.*

Definition 4.6 *Let C be a subset of $\mathbb{R}^J$. A point x in C is said to be an* interior point *of set C if there is $\epsilon > 0$ such that every point z with $||x - z|| < \epsilon$ is in C. The* interior *of the set C, written int(C), is the set of all interior points of C. It is also the largest open set contained within C.*

For example, the open interval $(0, 1)$ is the interior of the intervals $(0, 1]$ and $[0, 1]$. A set C is open if and only if $C = \text{int}(C)$.

Definition 4.7 *A point x in $\mathbb{R}^J$ is said to be a* boundary point *of set C if, for every $\epsilon > 0$, there are points y_ϵ in C and z_ϵ not in C, both depending on the choice of ϵ, with $||x - y_\epsilon|| < \epsilon$ and $||x - z_\epsilon|| < \epsilon$. The* boundary *of C is the set of all boundary points of C. It is also the intersection of the closure of C with the closure of its complement.*

For example, the points $x = 0$ and $x = 1$ are boundary points of the set $(0, 1]$.

Definition 4.8 *For $k = 0, 1, 2, ...,$ let x^k be a vector in $\mathbb{R}^J$. The sequence of vectors $\{x^k\}$ is said to* converge *to the vector z if, given any $\epsilon > 0$, there is positive integer n, usually depending on ϵ, such that, for every $k > n$, we have $||z - x^k|| \leq \epsilon$. Then we say that z is the* limit of the sequence.

For example, the sequence $\{x^k = \frac{1}{k+1}\}$ in R converges to $z = 0$. The sequence $\{(-1)^k\}$ alternates between 1 and -1, so does not converge. However, the subsequence associated with odd k converges to $z = -1$, while the subsequence associated with even k converges to $z = 1$. The values $z = -1$ and $z = 1$ are called *subsequential limit points*, or, sometimes, *cluster points* of the sequence.

Definition 4.9 *A sequence $\{x^k\}$ of vectors in $\mathbb{R}^J$ is said to be* bounded *if there is a constant $b > 0$, such that $||x^k|| \leq b$, for all k.*

A fundamental result in analysis is the following.

Proposition 4.1 *Every convergent sequence of vectors in $\mathbb{R}^J$ is bounded. Every bounded sequence of vectors in $\mathbb{R}^J$ has at least one convergent subsequence, therefore, has at least one cluster point.*

4.4 Convex Sets in $\mathbb{R}^J$

In preparation for our discussion of linear and nonlinear programming, we consider some of the basic concepts from the geometry of convex sets.

4.4.1 Basic Definitions

We begin with the basic definitions.

Definition 4.10 *A vector z is said to be a* convex combination *of the vectors x and y if there is α in the interval $[0, 1]$ such that $z = (1-\alpha)x+\alpha y$. More generally, a vector z is a convex combination of the vectors x^n, $n = 1, ..., N$, if there are numbers $\alpha_n \geq 0$ with*

$$\alpha_1 + ... + \alpha_N = 1$$

and

$$z = \alpha_1 x^1 + ... + \alpha_N x^N.$$

Definition 4.11 *A nonempty set C in $\mathbb{R}^J$ is said to be* convex *if, for any distinct points x and y in C, and for any real number α in the interval $(0, 1)$, the point $(1 - \alpha)x + \alpha y$ is also in C; that is, C is closed to convex combinations of any two members of C.*

In Exercise 4.2 you are asked to show that if C is convex then the convex combination of any number of members of C is again in C. We say then that C is *closed to convex combinations.*

For example, the two-norm unit ball B in $\mathbb{R}^J$, consisting of all x with $||x||_2 \leq 1$, is convex, while the surface of the ball, the set of all x with $||x||_2 = 1$, is not convex. More generally, the unit ball of $\mathbb{R}^J$ in any norm is a convex set, as a consequence of the triangle inequality for norms.

Definition 4.12 *The* convex hull *of a set S, denoted conv(S), is the smallest convex set containing S, by which we mean that if K is any convex set containing S, then K must also contain conv(S).*

One weakness of this definition is that it does not tell us explicitly what the members of conv(S) look like, nor precisely how the individual members of conv(S) are related to the members of S itself. In fact, it is not obvious that a smallest such set exists at all. The following proposition remedies this; the reader is asked to supply a proof in Exercise 4.3 later.

Proposition 4.2 *The convex hull of a set S is the set C of all convex combinations of members of S.*

Definition 4.13 *A subset S of $\mathbb{R}^J$ is a* subspace *if, for every x and y in S and scalars α and β, the linear combination $\alpha x + \beta y$ is again in S.*

A subspace is necessarily a convex set.

Definition 4.14 *The* orthogonal complement *of a subspace S of $\mathbb{R}^J$, endowed with the two-norm, is the set*

$$S^\perp = \{u | \langle u, s\rangle = u \cdot s = u^T s = 0, \text{for every } s \in S\},$$

the set of all vectors u in $\mathbb{R}^J$ that are orthogonal to every member of S.

For example, in $\mathbb{R}^3$, the x, y-plane is a subspace and has for its orthogonal complement the z-axis.

Definition 4.15 *A subset M of $\mathbb{R}^J$ is a* linear manifold *if there is a subspace S and a vector b such that*

$$M = S + b = \{x | x = s + b, \text{for some } s \text{ in } S\}.$$

Any linear manifold is convex.

Definition 4.16 *For a fixed column vector a with Euclidean length one and a fixed scalar γ the* hyperplane *determined by a and γ is the set*

$$H(a, \gamma) = \{z | \langle a, z\rangle = \gamma\}.$$

The hyperplanes $H(a, \gamma)$ are linear manifolds, and the hyperplanes $H(a, 0)$ are subspaces. Hyperplanes in $\mathbb{R}^J$ are naturally associated with linear equations in J variables; with $a = (a_1, ..., a_J)^T$, the hyperplane $H(a, \gamma)$ is the set of all $z = (z_1, ..., z_J)^T$ for which

$$a_1 z_1 + a_2 z_2 + ... + a_J z_J = \gamma.$$

Earlier, we mentioned that there are two related, but distinct, ways to view members of the set $\mathbb{R}^J$. The first is to see x in $\mathbb{R}^J$ as a point in J-dimensional space, so that, for example, if $J = 2$, then a member x of $\mathbb{R}^2$ can be thought of as a point in a plane, the plane of the blackboard, say. The second way is to think of x as the directed line segment from the origin to the point also denoted x. We purposely avoided making a choice between one interpretation and the other because there are cases in which we want to employ both interpretations; the definition of the hyperplane $H(a, \gamma)$ provides just such a case. We want to think of the members of the hyperplane as points in $\mathbb{R}^J$ that lie within the set $H(a, \gamma)$, but we want to think of a as a directed line segment perpendicular, or normal, to the hyperplane. When x, viewed as a point, is in $H(a, \gamma)$, the directed line segment from the origin to x will not lie in the hyperplane, unless $\gamma = 0$.

Lemma 4.1 *The distance from the hyperplane $H(a, \gamma)$ to the hyperplane $H(a, \gamma + 1)$ is one.*

The proof is left as Exercise 4.9.

Definition 4.17 *For each vector a with $\|a\|_2 = 1$ and each scalar γ, the sets*

$$H_+(a, \gamma) = \{z | \langle a, z \rangle \geq \gamma\}$$

$$H_-(a, \gamma) = \{z | \langle a, z \rangle \leq \gamma\}$$

are half-spaces.

Half-spaces in $\mathbb{R}^J$ are naturally associated with linear inequalities in J variables; with $a = (a_1, ..., a_J)^T$, the half-space $H_+(a, \gamma)$ is the set of all $z = (z_1, ..., z_J)^T$ for which

$$a_1 z_1 + a_2 z_2 + ... + a_J z_J \geq \gamma.$$

Perhaps the most important convex sets in optimization are the *polyhedrons*:

Definition 4.18 *A subset P of $\mathbb{R}^J$ is a* polyhedron *if P is the intersection of a finite number of half-spaces.*

A polyhedron is the set of all vectors that satisfy a finite number of linear inequalities: the set P in $\mathbb{R}^2$ consisting of all vectors (x_1, x_2) with $x_1 \geq 0$, $x_2 \geq 0$ is an *unbounded* polyhedron, while the set B in $\mathbb{R}^2$ consisting of all vectors (x_1, x_2) with $x_1 \geq 0$, $x_2 \geq 0$ and $x_1 + x_2 \leq 1$ is a *bounded* polyhedron. The set B is also the convex hull of a finite set of points, namely the three points $(0,0), (1,0)$ and $(0,1)$, and therefore is also a *polytope.*

Definition 4.19 *Given a subset C of $\mathbb{R}^J$, the* affine hull *of C, denoted aff(C), is the smallest linear manifold containing C.*

For example, let C be the line segment connecting the two points $(0,1)$ and $(1,2)$ in $\mathbb{R}^2$. The affine hull of C is the straight line whose equation is $y = x + 1$.

Definition 4.20 *The* dimension *of a subset of $\mathbb{R}^J$ is the dimension of its affine hull, which is the dimension of the subspace of which it is a translate.*

The set C above has dimension one. A set containing only one point is its own affine hull, since it is a translate of the subspace $\{0\}$.

In $\mathbb{R}^2$, the line segment connecting the points $(0,1)$ and $(1,2)$ has no interior; it is a one-dimensional subset of a two-dimensional space and can contain no two-dimensional ball. But, the part of this set without its two end points is a sort of interior, called the *relative interior.*

Definition 4.21 *The* relative interior *of a subset C of $\mathbb{R}^J$, denoted ri(C), is the interior of C, as defined by considering C as a subset of its affine hull.*

Since a set consisting of a single point is its own affine hull, it is its own relative interior.

Definition 4.22 *A point x in a convex set C is said to be an* extreme point *of C if the set obtained by removing x from C remains convex.*

Said another way, $x \in C$ is an extreme point of C if x is not a convex combination of two other points in C; that is, x cannot be written as

$$x = (1 - \alpha)y + \alpha z,$$

for y and z in C, $y, z \neq x$ and $\alpha \in (0,1)$. For example, the point $x = 1$ is an extreme point of the convex set $C = [0,1]$. Every point on the boundary of a sphere in $\mathbb{R}^J$ is an extreme point of the sphere. The set of all extreme points of a convex set is denoted Ext(C).

Definition 4.23 *A non-zero vector d is said to be a* direction of unboundedness *of a convex set C if, for all x in C and all $\gamma \geq 0$, the vector $x + \gamma d$ is in C.*

For example, if C is the nonnegative orthant in $\mathbb{R}^J$, then any nonnegative vector d is a direction of unboundedness.

Definition 4.24 *A vector a is* normal *to a convex set C at the point s in C if*

$$\langle a, c - s \rangle \leq 0,$$

for all c in C.

Definition 4.25 *Let C be convex and s in C. The* normal cone *to C at s, denoted $N_C(s)$, is the set of all vectors a that are normal to C at s.*

Normality and the normal cone are notions that make sense only in a space with an inner product, so are implicitly connected to the two-norm.

4.4.2 Orthogonal Projection onto Convex Sets

The following proposition is fundamental in the study of convexity and can be found in most books on the subject; see, for example, the text by Goebel and Reich [118].

Proposition 4.3 *Given any nonempty closed convex set C and an arbitrary vector x in $\mathbb{R}^J$, there is a unique member $P_C x$ of C closest, in the sense of the two-norm, to x. The vector $P_C x$ is called the* orthogonal (or metric) projection *of x onto C and the operator P_C the* orthogonal projection *onto C.*

Proof: If x is in C, then $P_C x = x$, so assume that x is not in C. Then $d > 0$, where d is the distance from x to C. For each positive integer n, select c^n in C with $||x - c^n||_2 < d + \frac{1}{n}$. Then, since for all n we have

$$||c^n||_2 = ||c^n - x + x||_2 \leq ||c^n - x||_2 + ||x||_2 \leq d + \frac{1}{n} + ||x||_2 < d + 1 + ||x||_2,$$

the sequence $\{c^n\}$ is bounded; let c^* be any cluster point. It follows easily that $||x - c^*||_2 = d$ and that c^* is in C. If there is any other member c of C with $||x - c||_2 = d$, then, by the Parallelogram Law, we would have $||x - (c^* + c)/2||_2 < d$, which is a contradiction. Therefore, c^* is $P_C x$. ∎

The proof just given relies on the Bolzano–Weierstrass Theorem 3.1. There is another proof, which avoids this theorem and so is valid for infinite-dimensional Hilbert space. The idea is to use the Parallelogram Law to show that the sequence $\{c^n\}$ is Cauchy and then to use completeness to get c^*. We leave the details to the reader.

Here are some examples of orthogonal projection. If $C = U$, the unit ball, then $P_C x = x/||x||_2$, for all x such that $||x||_2 > 1$, and $P_C x = x$ otherwise. If C is $\mathbb{R}^J_+$, the nonnegative cone of $\mathbb{R}^J$, consisting of all vectors

x with $x_j \geq 0$, for each j, then $P_C x = x_+$, the vector whose entries are $\max(x_j, 0)$. For any closed, convex set C, the distance from x to C is $||x - P_C x||_2$.

If a nonempty closed set S is not convex, then the orthogonal projection of a vector x onto S need not be well defined; there may be more than one vector in S closest to x. In fact, it is known that a closed set S is convex if and only if, for every x not in S, there is a unique point in S closest to x; this is Motzkin's Theorem (see [24], p. 447). Note that there may well be some x for which there is a unique closest point in S, but if S is closed, but not convex, then there must be at least one point without a unique closest point in S.

The main reason for not speaking about orthogonal projection in the context of other norms is that there need not be a unique closest point in C to x; remember that the Parallelogram Law need not hold. For example, consider the closed convex set C in $\mathbb{R}^2$ consisting of all vectors $(a, b)^T$ with $a \geq 0$, $b \geq 0$, and $a + b = 1$. Let $x = (1, 1)^T$. Then each point in C is a distance one from x, in the sense of the one-norm.

Lemma 4.2 *For $H = H(a, \gamma)$, $z = P_H x$ is the vector*

$$z = P_H x = x + (\gamma - \langle a, x \rangle)a.$$

Proof: In Exercise 4.10 the reader is asked to supply a proof. ∎

We shall use this fact in our discussion of the ART algorithm.

For an arbitrary nonempty closed convex set C in $\mathbb{R}^J$, the orthogonal projection $T = P_C$ is a nonlinear operator, unless, of course, C is a subspace. We may not be able to describe $P_C x$ explicitly, but we do know a useful property of $P_C x$.

Proposition 4.4 *For a given x, a vector z in C is $P_C x$ if and only if*

$$\langle c - z, z - x \rangle \geq 0,$$

for all c in the set C.

Proof: Let c be arbitrary in C and α in $(0, 1)$. Then

$$||x - P_C x||_2^2 \leq ||x - (1 - \alpha)P_C x - \alpha c||_2^2 = ||x - P_C x + \alpha(P_C x - c)||_2^2$$
$$= ||x - P_C x||_2^2 - 2\alpha\langle x - P_C x, c - P_C x\rangle + \alpha^2 ||P_C x - c||_2^2.$$

Therefore,

$$-2\alpha\langle x - P_C x, c - P_C x\rangle + \alpha^2||P_C x - c||_2^2 \geq 0,$$

so that

$$2\langle x - P_C x, c - P_C x\rangle \leq \alpha||P_C x - c||_2^2.$$

Taking the limit, as $\alpha \to 0$, we conclude that

$$\langle c - P_C x, P_C x - x \rangle \geq 0.$$

If z is a member of C that also has the property

$$\langle c - z, z - x \rangle \geq 0,$$

for all c in C, then we have both

$$\langle z - P_C x, P_C x - x \rangle \geq 0,$$

and

$$\langle z - P_C x, x - z \rangle \geq 0.$$

Adding on both sides of these two inequalities leads to

$$\langle z - P_C x, P_C x - z \rangle \geq 0.$$

But,

$$\langle z - P_C x, P_C x - z \rangle = -||z - P_C x||_2^2,$$

so it must be the case that $z = P_C x$. This completes the proof. ∎

Corollary 4.1 *For any x and y in $\mathbb{R}^J$ we have*

$$\langle P_C x - P_C y, x - y \rangle \geq \|P_C x - P_C y\|_2^2, \tag{4.4}$$

and so

$$\|P_C x - P_C y\|_2 \leq \|x - y\|_2;$$

that is, the operator P_C is nonexpansive.

Proof: Use Proposition 4.4 to get

$$\langle P_C y - P_C x, P_C x - x \rangle \geq 0,$$

or

$$\langle P_C x - P_C y, x - P_C x \rangle \geq 0, \tag{4.5}$$

and

$$\langle P_C x - P_C y, P_C y - y \rangle \geq 0. \tag{4.6}$$

Now add the two inequalities in (4.5) and (4.6) to obtain

$$\langle P_C x - P_C y, x - y \rangle \geq ||P_C x - P_C y||_2^2.$$

So P_C is firmly nonexpansive. ∎

Corollary 4.2 *If $\|P_C x - P_C y\| = \|x - y\|$, then $P_C x - x = P_C y - y$.*

Proof: The proof is left as Exercise 4.18. ∎

4.5 More on Projections

The characterization of the orthogonal-projection operator P_C given by Proposition 4.4 has a number of important consequences.

Corollary 4.3 *Let S be any subspace of $\mathbb{R}^J$. Then, for any x in $\mathbb{R}^J$ and s in S, we have*

$$\langle P_S x - x, s \rangle = 0.$$

Proof: Since S is a subspace, $s + P_S x$ is again in S, for all s, as is γs, for every scalar γ. ∎

This corollary enables us to prove the Decomposition Theorem.

Theorem 4.1 *Let S be any subspace of $\mathbb{R}^J$ and x any member of $\mathbb{R}^J$. Then there are unique vectors s in S and u in $S^\perp$ such that $x = s + u$. The vector s is $P_S x$ and the vector u is $P_{S^\perp} x$.*

Proof: For the given x we take $s = P_S x$ and $u = x - P_S x$. Corollary 4.3 assures us that u is in $S^\perp$. Now we need to show that this decomposition is unique. To that end, suppose that we can write $x = s_1 + u_1$, with s_1 in S and u_1 in $S^\perp$. Then Proposition 4.4 tells us that, since $s_1 - x$ is orthogonal to every member of S, s_1 must be $P_S x$. ∎

This theorem is often presented in a slightly different manner.

Theorem 4.2 *Let A be a real I by J matrix. Then every vector b in $\mathbb{R}^I$ can be written uniquely as $b = Ax + w$, where $A^T w = 0$.*

To derive Theorem 4.2 from Theorem 4.1, we simply let $S = \{Ax | x \in \mathbb{R}^J\}$. Then $S^\perp$ is the set of all w such that $A^T w = 0$. It follows that w is the member of the null space of A^T closest to b.

Here are additional consequences of Proposition 4.4.

Corollary 4.4 *Let S be any subspace of $\mathbb{R}^J$, d a fixed vector, and M the linear manifold $M = S + d = \{v = s + d | s \in S\}$, obtained by translating the members of S by the vector d. Then, for every x in $\mathbb{R}^J$ and every v in M, we have*

$$\langle P_M x - x, v - P_M x \rangle = 0.$$

Proof: Since v and $P_M x$ are in M, they have the form $v = s + d$, and $P_M x = \hat{s} + d$, for some s and $\hat{s}$ in S. Then $v - P_M x = s - \hat{s}$. ∎

Corollary 4.5 *Let H be the hyperplane $H(a, \gamma)$. Then, for every x, and every h in H, we have*

$$\langle P_H x - x, h - P_H x \rangle = 0.$$

Corollary 4.6 *Let S be a subspace of $\mathbb{R}^J$. Then $(S^\perp)^\perp = S$.*

Proof: We know from Theorem 4.1 that any x can be written as $x = s + u$, where s is in S and u is in $S^\perp$. Suppose now that x is in the set $(S^\perp)^\perp$. Then

$$\|x\|_2^2 = \langle x, x\rangle = \langle x, s+u\rangle = \langle x, s\rangle = \langle s+u, s\rangle = \langle s, s\rangle = \|s\|_2^2.$$

But we also have

$$\|x\|_2^2 = \|s+u\|_2^2 = \|s\|_2^2 + \|u\|_2^2 + 2\langle s, u\rangle = \|s\|_2^2 + \|u\|_2^2,$$

from which we conclude that $\|u\|_2^2 = 0$, and $x = s$ is in S. ∎

4.6 Linear and Affine Operators on $\mathbb{R}^J$

If A is a J by J real matrix, then we can define an operator T by setting $Tx = Ax$, for each x in $\mathbb{R}^J$; here Ax denotes the multiplication of the matrix A and the column vector x.

Definition 4.26 *An operator T is said to be a* linear operator *if*

$$T(\alpha x + \beta y) = \alpha Tx + \beta Ty,$$

for each pair of vectors x and y and each pair of scalars α and β.

Any operator T that comes from matrix multiplication, that is, for which $Tx = Ax$, is linear. In fact, all linear operators are of this type.

Lemma 4.3 *For $H = H(a, \gamma)$, $H_0 = H(a, 0)$, and any x and y in $\mathbb{R}^J$, we have*

$$P_H(x+y) = P_Hx + P_Hy - P_H0,$$

so that

$$P_{H_0}(x+y) = P_{H_0}x + P_{H_0}y,$$

that is, the operator P_{H_0} is an additive operator. In addition,

$$P_{H_0}(\alpha x) = \alpha P_{H_0}x,$$

so that P_{H_0} is a linear operator.

Definition 4.27 *If B is a J by J real matrix and d is a fixed nonzero vector in $\mathbb{R}^J$, the operator defined by $Tx = Bx + d$ is an* affine-linear operator *or just an* affine operator.

Lemma 4.4 *For any hyperplane* $H = H(a, \gamma)$ *and* $H_0 = H(a, 0)$,

$$P_H x = P_{H_0} x + P_H 0,$$

so P_H *is an affine-linear operator.*

Lemma 4.5 *For* $i = 1, ..., I$ *let* H_i *be the hyperplane* $H_i = H(a^i, \gamma_i)$, $H_{i0} = H(a^i, 0)$, *and* P_i *and* P_{i0} *the orthogonal projections onto* H_i *and* H_{i0}, *respectively. Let* T *be the operator* $T = P_I P_{I-1} \cdots P_2 P_1$. *Then* $Tx = Bx + d$, *for some square matrix* B *and vector* d; *that is,* T *is an affine-linear operator.*

4.7 The Fundamental Theorems

The Separation Theorem and the Support Theorem provide the foundation for the geometric approach to the calculus of functions of several variables.

A real-valued function $f(x)$ defined for real x has a derivative at $x = x_0$ if and only if there is a unique line through the point $(x_0, f(x_0))$ tangent to the graph of $f(x)$ at that point. If $f(x)$ is not differentiable at x_0, there may be more than one such tangent line, as happens with the function $f(x) = |x|$ at $x_0 = 0$. For functions of several variables the geometric view of differentiation involves tangent hyperplanes.

4.7.1 Basic Definitions

It is convenient for us to consider functions on $\mathbb{R}^J$ whose values may be infinite. For example, we define the *indicator function* of a set $C \subseteq \mathbb{R}^J$ to have the value zero for x in C, and the value $+\infty$ for x outside the set C.

Definition 4.28 *A function* $f : \mathbb{R}^J \to [-\infty, \infty]$ *is* proper *if there is no* x *for which* $f(x) = -\infty$ *and some* x *for which* $f(x) < +\infty$.

All the functions we shall consider in this text will be proper.

Definition 4.29 *Let* f *be a proper function defined on* $\mathbb{R}^J$. *The subset of* $\mathbb{R}^{J+1}$ *defined by*

$$\text{epi}(f) = \{(x, \gamma) | f(x) \leq \gamma\}$$

is the epigraph *of* f. *Then we say that* f *is* convex *if its epigraph is a convex set.*

Alternative definitions of convex function are presented in the exercises.

Definition 4.30 *The* effective domain *of a proper function* $f : \mathbb{R}^J \rightarrow (-\infty, \infty]$ *is the set*

$$\mathrm{dom}(f) = \{x | \, f(x) < +\infty\}.$$

It is also the projection onto $\mathbb{R}^J$ *of its epigraph.*

It is easily shown that the effective domain of a convex function is a convex set.

The important role played by hyperplanes tangent to the epigraph of f motivates our study of the relationship between hyperplanes and convex sets.

4.7.2 The Separation Theorem

The Separation Theorem, sometimes called the Geometric Hahn–Banach Theorem, is an easy consequence of the existence of orthogonal projections onto closed convex sets.

Theorem 4.3 (The Separation Theorem) *Let C be a closed nonempty convex set in $\mathbb{R}^J$ and x a point not in C. Then there is non-zero vector a in $\mathbb{R}^J$ and real number α such that*

$$\langle a, c \rangle \leq \alpha < \langle a, x \rangle,$$

for every c in C.

Proof: Let $z = P_C x$, $a = x - z$, and $\alpha = \langle a, z \rangle$. Then using Proposition 4.4, we have

$$\langle -a, c - z \rangle \geq 0,$$

or, equivalently,

$$\langle a, c \rangle \leq \langle a, z \rangle = \alpha \,,$$

for all c in C. But, we also have

$$\langle a, x \rangle = \langle a, x - z \rangle + \langle a, z \rangle = ||x - z||_2^2 + \alpha > \alpha.$$

This completes the proof. ∎

4.7.3 The Support Theorem

The Separation Theorem concerns a closed convex set C and a point x outside the set C, and asserts the existence of a hyperplane separating the two. Now we are concerned with a point z on the boundary of a convex set C, such as a point $(b, f(b))$ on the boundary of the epigraph of f.

The Support Theorem asserts the existence of a hyperplane through such a point z, having the convex set entirely contained in one of its half-spaces. If we knew a priori that the point z is $P_C x$ for some x outside C, then we could simply take the vector $a = x - z$ as the normal to the desired hyperplane. The essence of the Support Theorem is to provide such a normal vector without assuming that $z = P_C x$.

For the proofs that follow we shall need the following definitions.

Definition 4.31 *For subsets A and B of $\mathbb{R}^J$, and scalar γ, let the set $A + B$ consist of all vectors v of the form $v = a + b$, and γA consist of all vectors w of the form $w = \gamma a$, for some a in A and b in B. Let x be a fixed member of $\mathbb{R}^J$. Then the set $x + A$ is the set of all vectors y such that $y = x + a$, for some a in A.*

Lemma 4.6 *Let B be the unit ball in $\mathbb{R}^J$, that is, B is the set of all vectors u with $||u||_2 \leq 1$. Let S be an arbitrary subset of $\mathbb{R}^J$. Then x is in the interior of S if and only if there is some $\epsilon > 0$ such that $x + \epsilon B \subseteq S$, and y is in the closure of S if and only if, for every $\epsilon > 0$, the set $y + \epsilon B$ has nonempty intersection with S.*

We begin with the *Accessibility Lemma.* Note that the relative interior of any nonempty convex set is always nonempty (see [182], Theorem 6.2).

Lemma 4.7 (The Accessibility Lemma) *Let C be a convex set. Let x be in the relative interior of C and y in the closure of C. Then, for all scalars α in the interval $[0, 1)$, the point $(1 - \alpha)x + \alpha y$ is in the relative interior of C.*

Proof: If the dimension of C is less than J, we can transform the problem into a space of smaller dimension. Therefore, without loss of generality, we can assume that the dimension of C is J, its affine hull is all of $\mathbb{R}^J$, and its relative interior is its interior. Let α be fixed, and $B = \{z| \, ||z||_2 \leq 1\}$. We have to show that there is some $\epsilon > 0$ such that the set $(1 - \alpha)x + \alpha y + \epsilon B$ is a subset of the set C. We know that y is in the set $C + \epsilon B$ for every $\epsilon > 0$, since y is in the closure of C. Therefore, for all $\epsilon > 0$ we have

$$(1 - \alpha)x + \alpha y + \epsilon B \subseteq (1 - \alpha)x + \alpha(C + \epsilon B) + \epsilon B$$

$$= (1 - \alpha)x + (1 + \alpha)\epsilon B + \alpha C$$

$$= (1 - \alpha)[x + \epsilon(1 + \alpha)(1 - \alpha)^{-1}B] + \alpha C.$$

Since x is in the interior of the set C, we know that

$$[x + \epsilon(1 + \alpha)(1 - \alpha)^{-1}B] \subseteq C,$$

for ϵ small enough. This completes the proof. ∎

Theorem 4.4 (The Support Theorem) *Let C be convex, and let z be on the boundary of C. Then there is a non-zero vector a in $\mathbb{R}^J$ with $\langle a, z\rangle \geq \langle a, c\rangle$, for all c in C.*

Proof: If the dimension of C is less than J, then every point of C is on the boundary of C. Let the affine hull of C be $M = S + b$. Then the set $C - b$ is contained in the subspace S, which, in turn, can be contained in a hyperplane through the origin, $H(a, 0)$. Then

$$\langle a, c\rangle = \langle a, b\rangle,$$

for all c in C. So we focus on the case in which the dimension of C is J, in which case the interior of C must be nonempty.

Let y be in the interior of C, and, for each $t > 1$, let $z_t = y + t(z - y)$. Note that z_t is not in the closure of C, for any $t > 1$, by the Accessibility Lemma, since z is not in the interior of C. By the Separation Theorem, there are vectors b_t such that

$$\langle b_t, c\rangle < \langle b_t, z_t\rangle,$$

for all c in C. For convenience, we assume that $||b_t||_2 = 1$, and that $\{t_k\}$ is a sequence with $t_k > 1$ and $\{t_k\} \to 1$, as $k \to \infty$. Let $a_k = b_{t_k}$. Then there is a subsequence of the $\{a_k\}$ converging to some a, with $||a||_2 = 1$, and

$$\langle a, c\rangle \leq \langle a, z\rangle,$$

for all c in C. This completes the proof. ∎

If we had known that there was a vector x not in C, such that $z = P_C x$, then we could have choosen $a = x - z$, as in the proof of the Separation Theorem. The point of the Support Theorem is that we cannot assume, a priori, that there is such an x. Once we have the vector a, however, any point $x = z + \lambda a$, for $\lambda \geq 0$, has the property that $z = P_C x$.

4.8 Block-Matrix Notation

Beginning in the next section we shall make extensive use of what is called *block-matrix notation*. The following example will illustrate this concept. Consider the matrix multiplication

$$\begin{bmatrix} 1 & 2 & 3 \\ 4 & 5 & 6 \end{bmatrix} \begin{bmatrix} x \\ y \\ z \end{bmatrix} = \begin{bmatrix} x + 2y + 3z \\ 4x + 5y + 6z \end{bmatrix}.$$

We define

$$A = \begin{bmatrix} 1 & 2 \\ 4 & 5 \end{bmatrix}, B = \begin{bmatrix} 3 \\ 6 \end{bmatrix},$$

$$u = \begin{bmatrix} x \\ y \end{bmatrix}, \text{and } v = \begin{bmatrix} z \end{bmatrix}.$$

We then write

$$\begin{bmatrix} 1 & 2 & 3 \\ 4 & 5 & 6 \end{bmatrix} \begin{bmatrix} x \\ y \\ z \end{bmatrix} = \begin{bmatrix} A & B \end{bmatrix} \begin{bmatrix} u \\ v \end{bmatrix}.$$

We calculate the product using the rule

$$\begin{bmatrix} A & B \end{bmatrix} \begin{bmatrix} u \\ v \end{bmatrix} = Au + Bv,$$

just as if the A, u, B, and v were numbers. From

$$Au = \begin{bmatrix} 1 & 2 \\ 4 & 5 \end{bmatrix} \begin{bmatrix} x \\ y \end{bmatrix} = \begin{bmatrix} x + 2y \\ 4x + 5y \end{bmatrix}$$

and

$$Bv = \begin{bmatrix} 3 \\ 6 \end{bmatrix} \begin{bmatrix} z \end{bmatrix} = \begin{bmatrix} 3z \\ 6z \end{bmatrix},$$

we see that

$$Au + Bv = \begin{bmatrix} x + 2y + 3z \\ 4x + 5y + 6z \end{bmatrix},$$

which is what we got previously.

4.9 Theorems of the Alternative

In linear algebra the emphasis is on systems of linear equations; little time, if any, is spent on systems of linear inequalities. But linear inequalities are important in optimization. In this section we consider some of the basic theorems regarding linear inequalities. These theorems all fit a certain pattern, known as a *Theorem of the Alternative.* These theorems assert that precisely one of two problems will have a solution. The proof of the first theorem illustrates how we should go about proving such theorems.

Theorem 4.5 (Gale I)[115] *Precisely one of the following is true:*

(1) there is x such that $Ax = b$;

(2) there is y such that $A^T y = 0$ and $b^T y = 1$.

Proof: First, we show that it is not possible for both to be true at the same time. Suppose that $Ax = b$ and $A^T y = 0$. Then $b^T y = x^T A^T y = 0$, so that we cannot have $b^T y = 1$. By Theorem 4.1, the fundamental decomposition theorem from linear algebra, we know that, for any b, there are unique Ax and w with $A^T w = 0$ such that $b = Ax + w$. Clearly, $b = Ax$ if and only if $w = 0$. Also, $b^T y = w^T y$. Therefore, if alternative (1) does not hold, we must have w non-zero, in which case $A^T y = 0$ and $b^T y = 1$, for $y = w/||w||_2^2$, so alternative (2) holds. ∎

In this section we consider several other theorems of this type. Perhaps the most well known of these theorems of the alternative is Farkas' Lemma.

Theorem 4.6 (Farkas' Lemma)[110] *Precisely one of the following is true:*

(1) there is $x \geq 0$ such that $Ax = b$;

(2) there is y such that $A^T y \geq 0$ and $b^T y < 0$.

Proof: We can restate the lemma as follows. There is a vector y with $A^T y \geq 0$ and $b^T y < 0$ if and only if b is not a member of the convex set $C = \{Ax | x \geq 0\}$. If b is not in C, which is closed and convex, then, by the Separation Theorem, there is a non-zero vector a and real α with

$$\langle a, b \rangle = b^T a > \alpha \geq \langle a, Ax \rangle = (Ax)^T a = x^T A^T a,$$

for all $x \geq 0$. Since $x^T A^T a$ is bounded above, as x runs over all nonnegative vectors, it follows that $A^T a \leq 0$. Choosing $x = 0$, we have $\alpha \geq 0$. Then let $y = -a$. Conversely, if $Ax = b$ does have a nonnegative solution x, then $A^T y \geq 0$ implies that $y^T Ax = y^T b \geq 0$. ∎

The next theorem can be obtained from Farkas' Lemma.

Theorem 4.7 (Gale II)[115] *Precisely one of the following is true:*

(1) there is x such that $Ax \leq b$;

(2) there is $y \geq 0$ such that $A^T y = 0$ and $b^T y < 0$.

Proof: First, if both are true, then $0 \leq y^T(b - Ax) = y^T b - 0 = y^T b$, which is a contradiction. Now assume that (2) does not hold. Therefore, for every $y \geq 0$ with $A^T y = 0$, we have $b^T y \geq 0$. Let $B = \begin{bmatrix} A & b \end{bmatrix}$. Then the system $B^T y = \begin{bmatrix} 0 & -1 \end{bmatrix}^T$ has no nonnegative solution. Applying Farkas' Lemma, we find that there is a vector $w = \begin{bmatrix} z & \gamma \end{bmatrix}^T$ with $Bw \geq 0$ and $\begin{bmatrix} 0 & -1 \end{bmatrix} w < 0$. So, $Az + \gamma b \geq 0$ and $\gamma > 0$. Let $x = -\frac{1}{\gamma} z$ to get $Ax \leq b$, so that (1) holds. ∎

The next theorem also is a consequence of Farkas' Lemma.

Theorem 4.8 (Gordan)[120] *Precisely one of the following is true:*

(1) there is x such that $Ax < 0$;

(2) there is $y \geq 0$, $y \neq 0$, such that $A^T y = 0$.

Proof: First, if both are true, then $0 < -y^T Ax = 0$, which cannot be true. Now assume that there is no non-zero $y \geq 0$ with $A^T y = 0$. Then, with $e = (1, 1, ..., 1)^T$, $C = \begin{bmatrix} A & e \end{bmatrix}$, and $d = (0, 0, ..., 0, 1)^T$, there is no nonnegative solution of $C^T y = d$. From Farkas' Lemma we then know that there is a vector $z = \begin{bmatrix} u^T & \gamma \end{bmatrix}^T$, with $Cz = Au + \gamma e \geq 0$, and $d^T z < 0$. Then $Ax < 0$ for $x = -u$. ∎

Here are several more theorems of the alternative.

Theorem 4.9 (Stiemke I)[194] *Precisely one of the following is true:*

(1) there is x such that $Ax \leq 0$ and $Ax \neq 0$;

(2) there is $y > 0$ such that $A^T y = 0$.

Theorem 4.10 (Stiemke II)[194] *Let c be a fixed non-zero vector. Precisely one of the following is true:*

(1) there is x such that $Ax \leq 0$ and $c^T x \geq 0$ and not both $Ax = 0$ and $c^T x = 0$;

(2) there is $y > 0$ such that $A^T y = c$.

When we study linear programming in Chapter 6 we shall encounter David Gale's Strong Duality Theorem. His proof of that theorem will depend heavily on the following theorem of the alternative.

Theorem 4.11 (Gale III)[115] *Let b be a fixed non-zero vector. Precisely one of the following is true:*

(1) there is $x \geq 0$ such that $Ax \leq b$;

(2) there is $y \geq 0$ such that $A^T y \geq 0$ and $b^T y < 0$.

Proof: First, note that we cannot have both true at the same time, because $b^T y < 0$, $y \geq 0$, and $Ax \leq b$ would imply that $x^T A^T y = x \cdot A^T y < 0$, which is a contradiction. Now suppose that (1) does not hold. Then there is no $w = \begin{bmatrix} x \\ u \end{bmatrix} \geq 0$ such that

$$\begin{bmatrix} A & I \end{bmatrix} w = b.$$

By Farkas' Lemma (Theorem 4.6), it follows that there is y with

$$\begin{bmatrix} A^T \\ I \end{bmatrix} y \geq 0,$$

and $b^T y < 0$. Therefore, $A^T y \geq 0$, $Iy = y \geq 0$, and $b^T y < 0$; therefore, (2) holds. ∎

Theorem 4.12 (von Neumann)[167] *Precisely one of the following is true:*

- *(1) there is $x \geq 0$ such that $Ax > 0$;*
- *(2) there is $y \geq 0$, $y \neq 0$, such that $A^T y \leq 0$.*

Proof: If both were true, then we would have

$$0 < (Ax)^T y = x^T (A^T y),$$

so that $A^T y \leq 0$ would be false. Now suppose that (2) does not hold. Then there is no $y \geq 0$, $y \neq 0$, with $A^T y \leq 0$. Consequently, there is no $y \geq 0$, $y \neq 0$, such that

$$\begin{bmatrix} A^T \\ -u^T \end{bmatrix} y = \begin{bmatrix} A^T y \\ -u^T y \end{bmatrix} \leq \begin{bmatrix} 0 \\ -1 \end{bmatrix},$$

where $u^T = (1, 1, ..., 1)$. By Theorem 4.11, there is

$$z = \begin{bmatrix} x \\ \alpha \end{bmatrix} \geq 0,$$

such that

$$\begin{bmatrix} A & -u \end{bmatrix} z = \begin{bmatrix} A & -u \end{bmatrix} \begin{bmatrix} x \\ \alpha \end{bmatrix} \geq 0,$$

and

$$\begin{bmatrix} 0^T & -1 \end{bmatrix} z = \begin{bmatrix} 0^T & -1 \end{bmatrix} \begin{bmatrix} x \\ \alpha \end{bmatrix} = -\alpha < 0.$$

Therefore, $\alpha > 0$ and $(Ax)_i - \alpha \geq 0$ for each i, and so $Ax > 0$ and (1) holds. ∎

Theorem 4.13 (Tucker)[197] *Precisely one of the following is true:*

- *(1) there is $x \geq 0$ such that $Ax \geq 0$, $Ax \neq 0$;*
- *(2) there is $y > 0$ such that $A^T y \leq 0$.*

Theorem 4.14 (Theorem 21.1, [182]) *Let C be a convex set, and let $f_1, ..., f_m$ be proper convex functions, with $ri(C) \subseteq dom(f_i)$, for each i. Precisely one of the following is true:*

- *(1) there is $x \in C$ such that $f_i(x) < 0$, for $i = 1, ..., m$;*

(2) there are $\lambda_i \geq 0$*, not all equal to zero, such that*

$$\lambda_1 f_1(x) + ... + \lambda_m f_m(x) \geq 0,$$

for all x *in* C*.*

Theorem 4.14 is fundamental in proving Helly's Theorem.

Theorem 4.15 (Helly's Theorem) [182] *Let* $\{C_i \,|i = 1, ..., I\}$ *be a finite collection of (not necessarily closed) convex sets in* $\mathbb{R}^N$*. If every subcollection of* $N+1$ *or fewer sets has nonempty intersection, then the entire collection has nonempty intersection.*

For instance, in the two-dimensional plane, if a finite collection of lines is such that every three have a common point of intersection, then they all have a common point of intersection. There is another version of Helly's Theorem that applies to convex inequalities.

Theorem 4.16 *Let there be given a system of the form*

$$f_1(x) < 0, ..., f_k(x) < 0, f_{k+1}(x) \leq 0, ..., f_m(x) \leq 0,$$

where the f_i *are convex functions on* $\mathbb{R}^J$*, and the inequalities may be all strict or all weak. If every subsystem of* $J+1$ *or fewer inequalities has a solution in a given convex set* C*, then the entire system has a solution in* C*.*

4.10 Another Proof of Farkas' Lemma

In the previous section, we proved Farkas' Lemma, Theorem 4.6, using the Separation Theorem, the proof of which, in turn, depended here on the existence of the orthogonal projection onto any closed convex set. It is possible to prove Farkas' Lemma directly, along the lines of Gale [115].

Suppose that $Ax = b$ has no nonnegative solution. If, indeed, it has no solution whatsoever, then $b = Ax + w$, where $w \neq 0$ and $A^T w = 0$. Then we take $y = -w/||w||_2^2$. So suppose that $Ax = b$ does have solutions, but not any nonnegative ones. The approach is to use induction on the number of columns of the matrix involved in the lemma.

If A has only one column, denoted a^1, then $Ax = b$ can be written as

$$x_1 a^1 = b.$$

Assuming that there are no nonnegative solutions, it must follow that $x_1 < 0$. We take $y = -b$. Then

$$b^T y = -b^T b = -||b||_2^2 < 0,$$

while

$$A^T y = (a^1)^T(-b) = \frac{-1}{x_1} b^T b > 0.$$

Now assume that the lemma holds whenever the involved matrix has no more than $m-1$ columns. We show the same is true for m columns.

If there is no nonnegative solution of the system $Ax = b$, then clearly there are no nonnegative real numbers $x_1, x_2, ..., x_{m-1}$ such that

$$x_1 a^1 + x_2 a^2 + ... + x_{m-1} a^{m-1} = b,$$

where a^j denotes the jth column of the matrix A. By the induction hypothesis, there must be a vector v with

$$(a^j)^T v \geq 0,$$

for $j = 1, ..., m-1$, and $b^T v < 0$. If it happens that $(a^m)^T v \geq 0$ also, then we are done. If, on the other hand, we have $(a^m)^T v < 0$, then let

$$c^j = (a^j)^T a^m - (a^m)^T a^j,\ j = 1, ..., m-1,$$

and

$$d = (b^T v) a^m - ((a^m)^T v) b.$$

Then there are no nonnegative real numbers $z_1, ..., z_{m-1}$ such that

$$z_1 c^1 + z_2 c^2 + ... + z_{m-1} c^{m-1} = d, \tag{4.7}$$

since, otherwise, it would follow from simple calculations that

$$\frac{-1}{(a^m)^T v}\left(\left[\sum_{j=1}^{m-1} z_j((a^j)^T v)\right] - b^T v\right) a^m - \sum_{j=1}^{m-1} z_j((a^m)^T v) a^j = b.$$

Close inspection of this shows all the coefficients to be nonnegative, which implies that the system $Ax = b$ has a nonnegative solution, contrary to our assumption. It follows, therefore, that there can be no nonnegative solution to the system in Equation (4.7).

By the induction hypothesis, it follows that there is a vector u such that

$$(c^j)^T u \geq 0, \; j = 1, ..., m-1,$$

and

$$d^T u < 0.$$

Now let

$$y = ((a^m)^T u)v - ((a^m)^T v)u.$$

We can easily verify that

$$(a^j)^T y = (c^j)^T u \geq 0, \; j = 1, ..., m-1,$$

$$b^T y = d^T u < 0,$$

and

$$(a^m)^T y = 0,$$

so that

$$A^T y \geq 0,$$

and

$$b^T y < 0.$$

This completes the proof. ∎

4.11 Gordan's Theorem Revisited

In their text [23], Borwein and Lewis give the following version of Gordan's Theorem 4.8.

Theorem 4.17 *For any vectors $a^0, a^1, ..., a^m$ in $\mathbb{R}^J$, exactly one of the following systems has a solution:*

$$\sum_{i=0}^{m} \lambda_i a^i = 0, \; \sum_{i=0}^{m} \lambda_i = 1, \; 0 \leq \lambda_0, \lambda_1, ..., \lambda_m; \tag{4.8}$$

or there is some x for which

$$x^T a^i < 0, \; \text{for}\, i = 0, 1, ..., m. \tag{4.9}$$

The following proposition will be useful in the proof.

Proposition 4.5 *If the function $f : \mathbb{R}^J \to \mathbb{R}$ is differentiable and bounded below, that is, there is a constant α such that $\alpha \leq f(x)$ for all x, then for every $\epsilon > 0$ there is a point x^ϵ with $\|\nabla f(x^\epsilon)\|_2 \leq \epsilon$.*

Proof: Fix $\epsilon > 0$. The function $f(x) + \epsilon\|x\|_2$ has bounded level sets, so, by Proposition 3.2, it has a global minimizer, which we denote by x^ϵ. We show that $d = \nabla f(x^\epsilon)$ has $\|d\|_2 \leq \epsilon$.

If not, then $\|d\|_2 > \epsilon$. From the inequality

$$\lim_{t \downarrow 0} \frac{f(x^\epsilon - td) - f(x^\epsilon)}{t} = -\langle \nabla f(x^\epsilon), d\rangle = -\|d\|_2^2 < -\epsilon\|d\|_2$$

we would have, for small positive t,

$$-t\epsilon\|d\|_2 > f(x^\epsilon - td) - f(x^\epsilon)$$

$$= (f(x^\epsilon - td) + \epsilon\|x^\epsilon - td\|_2) - (f(x^\epsilon) + \epsilon\|x^\epsilon\|_2)$$

$$+\epsilon(\|x^\epsilon\|_2 - \|x^\epsilon - td\|_2) \geq -t\epsilon\|d\|_2,$$

which is impossible. ∎

Rather than prove Theorem 4.8 using the theory of convex sets and separation, as we did previously, they take the following approach. Let

$$f(x) = \log\Big(\sum_{i=0}^{m} \exp(x^T a^i)\Big).$$

We then have the following theorem.

Theorem 4.18 *The following statements are equivalent:*

(1) The function $f(x)$ is bounded below.

(2) System (4.8) is solvable.

(3) System (4.9) is unsolvable.

Proof: Showing that (2) implies (3) is easy. To show that (3) implies (1), note that if $f(x)$ is not bounded below, then there is some x with $f(x) \leq 0$, which forces $x^T a^i < 0$, for all i. Finally, to show that (1) implies (2), we use Proposition 4.5. Then there is a sequence $\{x^n\}$ with $\|\nabla f(x^n)\|_2 \leq \frac{1}{n}$, for each n. Since

$$\nabla f(x^n) = \sum_{i=0}^{m} \lambda_i^n a^i,$$

for

$$\lambda_i^n = \exp((x^n)^T a^i) / \sum_{i=0}^{m} \exp((x^n)^T a^i),$$

it follows that

$$\|\sum_{i=0}^{m} \lambda_i^n a^i\|_2 < \frac{1}{n},$$

for each n. The sequence $\{\lambda^n\}$ is bounded, so there is a convergent subsequence, converging to some λ^* for which $\sum_{i=0}^{m} \lambda_i^* a^i = 0$. ∎

4.12 Exercises

Ex. 4.1 *A more geometric approach to Cauchy's Inequality begins with an alternative approach to orthogonality [68]. Let x and y be nonzero vectors in $\mathbb{R}^J$. Say that x is orthogonal to y if*

$$\|x - y\|_2 = \|x + y\|_2.$$

To visualize this, draw a triangle with vertices x, y and $-y$. Then show that x and y are orthogonal if and only if $x \cdot y = 0$ and if and only if Pythagoras' Theorem holds; that is,

$$\|x - y\|_2^2 = \|x\|_2^2 + \|y\|_2^2.$$

Let p be the orthogonal projection of x on the line determined by y and the origin. Then $p = \gamma y$ for some constant γ. First, find γ using the fact that y and $x - \gamma y$ are orthogonal. Then use Pythagoras' Theorem to obtain Cauchy's Inequality. Define the angle between vectors x and y to be α such that

$$\cos \alpha = \frac{x \cdot y}{\|x\|_2 \|y\|_2}.$$

Then use this to prove the Law of Cosines and the Triangle Inequality.

Ex. 4.2 *Let $C \subseteq \mathbb{R}^J$, and let x^n, $n = 1, ..., N$ be members of C. For $n = 1, ..., N$, let $\alpha_n > 0$, with $\alpha_1 + ... + \alpha_N = 1$. Show that, if C is convex, then the* convex combination

$$\alpha_1 x^1 + \alpha_2 x^2 + ... + \alpha_N x^N$$

is in C.

Ex. 4.3 *Prove Proposition 4.2. Hint: Show that the set C is convex.*

Ex. 4.4 *Show that the subset of $\mathbb{R}^J$ consisting of all vectors x with $\|x\|_2 = 1$ is not convex.*

Ex. 4.5 *Let $\|x\|_2 = \|y\|_2 = 1$ and $z = \frac{1}{2}(x+y)$ in $\mathbb{R}^J$. Show that $\|z\|_2 < 1$ unless $x = y$. Show that this conclusion does not hold if the* two-norm $\|\cdot\|_2$ *is replaced by the* one-norm, *defined by*

$$\|x\|_1 = \sum_{j=1}^{J} |x_j|.$$

Ex. 4.6 *Let C be the set of all vectors x in $\mathbb{R}^J$ with $\|x\|_2 \leq 1$. Let K be a subset of C obtained by removing from C any number of its members for which $\|x\|_2 = 1$. Show that K is convex. Consequently, every x in C with $\|x\|_2 = 1$ is an extreme point of C.*

Ex. 4.7 *Prove that every subspace of $\mathbb{R}^J$ is convex, and every linear manifold is convex.*

Ex. 4.8 *Prove that every hyperplane $H(a, \gamma)$ is a linear manifold.*

Ex. 4.9 *Prove Lemma 4.1.*

Ex. 4.10 *Prove Lemma 4.2.*

Ex. 4.11 *Let A and B be nonempty, closed convex subsets of $\mathbb{R}^J$. Define the set $B - A$ to be all x in $\mathbb{R}^J$ such that $x = b - a$ for some $a \in A$ and $b \in B$. Show that $B - A$ is closed if one of the two sets is bounded. Find an example of two disjoint unbounded closed convex sets in $\mathbb{R}^2$ that get arbitrarily close to each other. Show that, for this example, $B - A$ is not closed.*

Ex. 4.12 *(a) Let C be a circular region in $\mathbb{R}^2$. Determine the normal cone for a point on its circumference. (b) Let C be a rectangular region in $\mathbb{R}^2$. Determine the normal cone for a point on its boundary.*

Ex. 4.13 *Prove Lemmas 4.3, 4.4 and 4.5.*

Ex. 4.14 *Let C be a convex set and $f : C \subseteq \mathbb{R}^J \to (-\infty, \infty]$. Prove that $f(x)$ is a convex function, according to Definition 4.29, if and only if, for all x and y in C, and for all $0 < \alpha < 1$, we have*

$$f(\alpha x + (1-\alpha)y) \leq \alpha f(x) + (1-\alpha)f(y).$$

Ex. 4.15 *Let $f : \mathbb{R}^J \to [-\infty, \infty]$. Prove that $f(x)$ is a convex function if and only if, for all $0 < \alpha < 1$, we have*

$$f(\alpha x + (1-\alpha)y) < \alpha b + (1-\alpha)c,$$

whenever $f(x) < b$ and $f(y) < c$.

Ex. 4.16 *Show that the vector a is orthogonal to the hyperplane $H = H(a, \gamma)$; that is, if u and v are in H, then a is orthogonal to $u - v$.*

Ex. 4.17 *Given a point s in a convex set C, where are the points x for which $s = P_C x$?*

Ex. 4.18 *Prove Corollary 4.2.*

Ex. 4.19 *Show that it is possible to have a vector $z \in \mathbb{R}^J$ such that $\langle z - x, c - z \rangle \geq 0$ for all $c \in C$, but z is not $P_C x$.*

Ex. 4.20 *Let z and a be as in the Support Theorem, let $\gamma > 0$, and let $x = z + \gamma a$. Show that $z = P_C x$.*

Ex. 4.21 *Let C be a closed, nonempty convex set in $\mathbb{R}^J$ and x not in C. Show that the distance from x to C is equal to the maximum of the distances from x to any hyperplane that separates x from C. Hint: Draw a picture.*

Ex. 4.22 *Let C be a nonempty set in $\mathbb{R}^J$ that is closed and convex, x a vector not in C, and $d > 0$ the distance from x to C. Let*

$$\sigma_C(a) = \sup_{c \in C} \langle a, c \rangle \,,$$

the support function *of C. Note that $\sigma_C(a)$ may be infinite for some a, but not for all a. Show that*

$$d = \max_{||a|| \leq 1} \{ \langle a, x \rangle - \sigma_C(a) \}.$$

The point here is to turn a minimization problem into one involving only maximization. Try drawing a picture and using Lemma 4.1. Hints: Consider the unit vector $\frac{1}{d}(x - P_C x)$, and use Cauchy's Inequality and Proposition 4.4. Remember that $P_C x$ is in C, so that

$$\langle a, P_C x \rangle \leq \sigma_C(a).$$

Remark: If, in the definition of the support function, we take the vectors a to be unit vectors, with $a = (\cos\theta, \sin\theta)$, for $0 \leq \theta < 2\pi$, then we can define the function

$$f(\theta) = \sup_{(x,y) \in C} x\cos\theta + y\sin\theta.$$

In [155] Tom Marzetta considers this function, as well as related functions of θ, such as the radius of curvature function, and establishes relationships between the behavior of these functions and the convex set itself.

Ex. 4.23 [10] *Let A and B be nonempty closed convex subsets of $\mathbb{R}^J$. For each $a \in A$ define*

$$d(a, B) = \inf_{b \in B} \|a - b\|_2,$$

and then define

$$d(A, B) = \inf_{a \in A} d(a, B).$$

Let

$$E = \{a \in A | d(a, B) = d(A, B)\},$$

and

$$F = \{b \in B | d(b, A) = d(B, A)\};$$

assume that both E and F are not empty. The displacement vector *is $v = P_K(0)$, where K is the closure of the set $B - A$. For any transformation $T : \mathbb{R}^J \to \mathbb{R}^J$, denote by* Fix$(T)$ *the set of all $x \in \mathbb{R}^J$ such that $Tx = x$. Prove the following:*

(a) $\|v\|_2 = d(A, B)$;

(b) $E + v = F$;

(c) $E = \text{Fix}(P_A P_B) = A \cap (B - v)$;

(d) $F = \text{Fix}(P_B P_A) = B \cap (A + v)$;

(e) $P_B e = P_F e = e + v, \text{for all } e \in E$;

(f) $P_A f = P_E f = f - v, \text{for all } f \in F$.

Ex. 4.24 *Let A and B be nonempty closed convex subsets of $\mathbb{R}^J$, with the distance between A and B given by $d = \inf_{a \in A, b \in B} \|a - b\|_2$. Assume that z is a fixed point of the operator $T = P_A P_B$.*

(a) Show that $\|z - P_B z\|_2 = d$.

(b) Let $x^{k+1} = Tx^k$, for $k = 0, 1, \ldots$. Show that $\{\|z - x^k\|_2\}$ is a decreasing sequence, so that the sequence $\{x^k\}$ is bounded and has at least one cluster point, x^*.

(c) Show that $Tx^* = x^*$ and the sequence $\{x^k\}$ converges to x^*.

Ex. 4.25 (Rådström Cancellation [23])

(a) Show that, for any subset S of $\mathbb{R}^N$, we have $2S \subseteq S + S$, and $2S = S + S$ if S is convex.

(b) Find three finite subsets of $\mathbb{R}$, say A, B, and C, with A not contained in B, but with the property that $A + C \subseteq B + C$. Hint: Try to find an example where the set C is $C = \{-1, 0, 1\}$.

(c) *Show that, if* A *and* B *are convex in* $\mathbb{R}^N$, B *is closed, and* C *is bounded in* $\mathbb{R}^N$, *then* $A+C \subseteq B+C$ *implies that* $A \subseteq B$. *Hint: Note that, under these assumptions,* $2A+C = A+(A+C) \subseteq 2B+C$.

Chapter 5

Vector Spaces and Matrices

5.1 Chapter Summary

In preparation for our discussion of linear programming, we present a brief review of the fundamentals of matrix theory.

5.2 Vector Spaces

Linear algebra is the study of *vector spaces* and *linear transformations.* It is not simply the study of matrices, although matrix theory takes up most of linear algebra.

It is common in mathematics to consider abstraction, which is simply a means of talking about more than one thing at the same time. A vector space $\mathcal{V}$ is an abstract algebraic structure defined using axioms. There are many examples of vector spaces, such as the sets of real or complex numbers themselves, the set of all polynomials, the set of row or column vectors of a given dimension, the set of all infinite sequences of real or complex numbers, the set of all matrices of a given size, and so on. The beauty of an abstract approach is that we can talk about all of these, and much more, all at once, without being specific about which example we mean.

A vector space is a set whose members are called *vectors*, on which there are two algebraic operations, called *scalar multiplication* and *vector addition*. As in any axiomatic approach, these notions are intentionally abstract. A vector is defined to be a member of a vector space, nothing more. Scalars are a bit more concrete, in that scalars are almost always real or complex numbers, although sometimes, but not in this book, they are members of an unspecified finite field. The operations themselves are not explicitly defined, except to say that they behave according to certain axioms, such as associativity and distributivity.

If v is a member of a vector space $\mathcal{V}$ and α is a scalar, then we denote by αv the scalar multiplication of v by α. If w is also a member of $\mathcal{V}$, then we denote by $v+w$ the vector addition of v and w. The following properties serve to define a vector space, with u, v, and w denoting arbitrary members of $\mathcal{V}$ and α and β arbitrary scalars:

(1) $v + w = w + v$;

(2) $u + (v + w) = (u + v) + w$;

(3) there is a "zero vector," denoted 0, with $v + 0 = v$ for all v;

(4) for each v there is a vector $-v$ such that $v + (-v) = 0$;

(5) $1v = v$, for all v;

(6) $(\alpha\beta)v = \alpha(\beta v)$;

(7) $\alpha(v + w) = \alpha v + \alpha w$;

(8) $(\alpha + \beta)v = \alpha v + \beta v$.

In Exercise 5.1 the reader is asked to prove that the zero vector is unique, that each v has a unique additive inverse $-v$, and that $(-1)v = -v$, for all v.

If $u^1, ..., u^N$ are members of $\mathcal{V}$ and $c_1, ..., c_N$ are scalars, then the vector

$$x = c_1u^1 + c_2u^2 + ... + c_Nu^N$$

is called a *linear combination* of the vectors $u^1, ..., u^N$, with coefficients $c_1, ..., c_N$.

If $\mathcal{W}$ is a subset of a vector space $\mathcal{V}$, then $\mathcal{W}$ is called a *subspace* of $\mathcal{V}$ if $\mathcal{W}$ is also a vector space for the same operations. What this means is simply that when we perform scalar multiplication on a vector in $\mathcal{W}$, or when we add vectors in $\mathcal{W}$, we always get members of $\mathcal{W}$ back again. Another way to say this is that $\mathcal{W}$ is *closed to linear combinations.*

When we speak of subspaces of $\mathcal{V}$ we do not mean to exclude the case of $\mathcal{W} = \mathcal{V}$. Note that $\mathcal{V}$ is itself a subspace, but not a *proper subspace*, of $\mathcal{V}$. Every subspace must contain the zero vector, 0; the smallest subspace of $\mathcal{V}$ is the subspace containing only the zero vector, $\mathcal{W} = \{0\}$.

In the vector space $\mathcal{V} = \mathbb{R}^2$, the subset of all vectors whose entries sum to zero is a subspace, but the subset of all vectors whose entries sum to one is not a subspace.

We often refer to things like $\begin{bmatrix} 1 & 2 & 0 \end{bmatrix}$ as vectors, although they are but one example of a certain type of vector. For clarity, in this book we shall call such an object a *real row vector of dimension three* or a *real row three-vector*. Similarly, we shall call $\begin{bmatrix} 3i \\ -1 \\ 2+i \\ 6 \end{bmatrix}$ a *complex column vector of dimension four* or a *complex column four-vector.*

The transpose of the row vector $\begin{bmatrix} 1 & 2 & 0 \end{bmatrix}$ is the column vector $\begin{bmatrix} 1 \\ 2 \\ 0 \end{bmatrix}$ and the transpose of the column vector $\begin{bmatrix} 1 \\ 2 \\ 0 \end{bmatrix}$ is the row vector $\begin{bmatrix} 1 & 2 & 0 \end{bmatrix}$. For any vector x the transpose of x is denoted x^T.

The conjugate transpose of the column vector $\begin{bmatrix} 3i \\ -1 \\ 2+i \\ 6 \end{bmatrix}$ is the row vector $\begin{bmatrix} -3i & -1 & 2-i & 6 \end{bmatrix}$, and the conjugate transpose of the row vector $\begin{bmatrix} -3i & -1 & 2-i & 6 \end{bmatrix}$ is the column vector $\begin{bmatrix} 3i \\ -1 \\ 2+i \\ 6 \end{bmatrix}$. For any vector x the conjugate transpose of x is denoted $x^\dagger$. For notational convenience, whenever we refer to something like a real three-vector or a complex four-vector, we shall always mean that they are columns, rather than rows. The space of real (column) N-vectors will be denoted $\mathbb{R}^N$, while the space of complex (column) N vectors is $\mathbb{C}^N$.

Shortly after beginning a discussion of vector spaces, we arrive at the notion of the size or dimension of the vector space. A vector space can be finite dimensional or infinite dimensional. The spaces $\mathbb{R}^N$ and $\mathbb{C}^N$ have dimension N; not a big surprise. The vector spaces of all infinite sequences of real or complex numbers are infinite dimensional, as is the vector space of all real or complex polynomials. If we choose to go down the path of finite dimensionality, we very quickly find ourselves talking about matrices. If we go down the path of infinite dimensionality, we quickly begin to discuss convergence of infinite sequences and sums, and find that we need to introduce norms, which takes us into functional analysis and the study of Hilbert and Banach spaces. In this course we shall consider only the finite dimensional vector spaces, which means that we shall be talking mainly about matrices.

5.3 Basic Linear Algebra

In this section we discuss bases and dimension, systems of linear equations, Gaussian elimination, and the notions of basic and non-basic variables.

5.3.1 Bases and Dimension

The notions of a basis and of linear independence are fundamental in linear algebra. Let $\mathcal{V}$ be a vector space.

Definition 5.1 *A collection of vectors* $\{u^1, ..., u^N\}$ *in* $\mathcal{V}$ *is* linearly independent *if there is no choice of scalars* $\alpha_1, ..., \alpha_N$*, not all zero, such that*

$$0 = \alpha_1 u^1 + ... + \alpha_N u^N.$$

Definition 5.2 *The* span *of a collection of vectors* $\{u^1, ..., u^N\}$ *in* $\mathcal{V}$ *is the set of all vectors* x *that can be written as linear combinations of the* u^n*; that is, for which there are scalars* $c_1, ..., c_N$*, such that*

$$x = c_1 u^1 + ... + c_N u^N.$$

Definition 5.3 *A collection of vectors* $\{w^1, ..., w^N\}$ *in* $\mathcal{V}$ *is called a* spanning set *for a subspace* $\mathcal{S}$ *if* $\mathcal{S}$ *is their span.*

Definition 5.4 *A collection of vectors* $\{u^1, ..., u^N\}$ *in* $\mathcal{V}$ *is called a* basis *for a subspace* $\mathcal{S}$ *if the collection is linearly independent and* $\mathcal{S}$ *is their span.*

Suppose that $\mathcal{S}$ is a subspace of $\mathcal{V}$, that $\{w^1, ..., w^N\}$ is a spanning set for $\mathcal{S}$, and $\{u^1, ..., u^M\}$ is a linearly independent subset of $\mathcal{S}$. Beginning with w^1, we augment the set $\{u^1, ..., u^M\}$ with w^j if w^j is not in the span of the u^m and the w^k previously included. At the end of this process, we have a linearly independent spanning set, and therefore, a basis, for $\mathcal{S}$. (Why?) Similarly, beginning with w^1, we remove w^j from the set $\{w^1, ..., w^N\}$ if w^j is a linear combination of the w^k, $k = 1, ..., j-1$. In this way we obtain a linearly independent set that spans $\mathcal{S}$, hence another basis for $\mathcal{S}$. The following lemma will allow us to prove that all bases for a subspace $\mathcal{S}$ have the same number of elements.

Lemma 5.1 *Let $W = \{w^1, ..., w^N\}$ be a spanning set for a subspace $\mathcal{S}$ in $\mathbb{R}^I$, and $V = \{v^1, ..., v^M\}$ a linearly independent subset of $\mathcal{S}$. Then $M \leq N$.*

Proof: Suppose that $M > N$. Let $B_0 = \{w^1, ..., w^N\}$. To obtain the set B_1, form the set $C_1 = \{v^1, w^1, ..., w^N\}$ and remove the first member of C_1 that is a linear combination of members of C_1 that occur to its left in the listing; since v^1 has no members to its left, it is not removed. Since W is a spanning set, v^1 is a linear combination of the members of W, so that some member of W is a linear combination of v^1 and the members of W that precede it in the list; remove the first member of W for which this is true.

We note that the set B_1 is a spanning set for $\mathcal{S}$ and has N members. Having obtained the spanning set B_k, with N members and whose first k members are $v^k, ..., v^1$, we form the set $C_{k+1} = B_k \cup \{v^{k+1}\}$, listing the members so that the first $k+1$ of them are $\{v^{k+1}, v^k, ..., v^1\}$. To get the set B_{k+1} we remove the first member of C_{k+1} that is a linear combination of the members to its left; there must be one, since B_k is a spanning set, and so v^{k+1} is a linear combination of the members of B_k. Since the set V is linearly independent, the member removed is from the set W. Continuing in this fashion, we obtain a sequence of spanning sets $B_1, ..., B_N$, each with N members. The set B_N is $B_N = \{v^1, ..., v^N\}$ and v^{N+1} must then be a linear combination of the members of B_N, which contradicts the linear independence of V. ∎

Corollary 5.1 *Every basis for a subspace $\mathcal{S}$ has the same number of elements.*

Definition 5.5 *The* dimension *of a subspace $\mathcal{S}$ is the number of elements in any basis.*

5.3.2 The Rank of a Matrix

Let A be an I by J matrix and x a J by 1 column vector. The equation $Ax = b$ tells us that the vector b is a linear combination of the columns of the matrix A, with the entries of the vector x as the coefficients; that is,

$$b = x_1 a^1 + x_2 a^2 + ... + x_J a^J,$$

where a^j denotes the jth column of A. Similarly, when we write the product $C = AB$, we are saying that the kth column of C is a linear combination of the columns of A, with the entries of the kth column of B as coefficients. It will be helpful to keep this in mind when reading the proof of the next lemma.

Lemma 5.2 *For any matrix A, the maximum number of linearly independent rows equals the maximum number of linearly independent columns.*

Proof: Suppose that A is an I by J matrix, and that $K \leq J$ is the maximum number of linearly independent columns of A. Select K linearly independent columns of A and use them as the K columns of an I by K matrix U. Since every column of A must be a linear combination of these K selected ones, there is a K by J matrix M such that $A = UM$. From $A^T = M^T U^T$ we conclude that every column of A^T is a linear combination of the K columns of the matrix M^T. Therefore, there can be at most K linearly independent columns of A^T. ∎

Definition 5.6 *The* rank *of A is the maximum number of linearly independent rows or of linearly independent columns of A.*

Proposition 5.1 *The rank of $C = AB$ is not greater than the smaller of the rank of A and the rank of B.*

Proof: Every column of C is a linear combination of the columns of A, so the rank of C cannot exceed that of A. Since the rank of C^T is the same as that of C, the proof is complete. ∎

Definition 5.7 *We say that an M by N matrix A has* full rank *if its rank is as large as possible; that is, the rank of A is the smaller of the two numbers M and N.*

Definition 5.8 *The N by N* identity matrix, *denoted I, has all its main diagonal entries equal to 1 and all other entries equal to 0. A square matrix A is* invertible *if there is a matrix B such that $AB = BA = I$. Then B is the inverse of A and we write $B = A^{-1}$.*

Proposition 5.2 *Let A be a square matrix. If there are matrices B and C such that $AB = I$ and $CA = I$, then $B = C = A^{-1}$.*

Proof: From $AB = I$ we have $C = C(AB) = (CA)B = IB = B$. ∎

Proposition 5.3 *A square matrix A is invertible if and only if it has full rank.*

Proof: We leave the proof as Exercise 5.3. ∎

Corollary 5.2 *A square matrix A is invertible if and only if there is a matrix B such that AB is invertible.*

There are many other conditions that are equivalent to A being invertible; we list several of these in the next subsection.

5.3.3 The "Matrix Inversion Theorem"

In this subsection we bring together several of the conditions equivalent to saying that an N by N matrix A is invertible. Taken together, these conditions are sometimes called the "Matrix Inversion Theorem". The equivalences on the list are roughly in increasing order of difficulty of proof. The reader is invited to supply proofs. We begin with the definition of invertibility.

(1) We say A is *invertible* if there is a matrix B such that $AB = BA = I$. Then $B = A^{-1}$, *the inverse* of A.

(2) A is invertible if and only if there are matrices B and C such that $AB = CA = I$. Then $B = C = A^{-1}$.

(3) A is invertible if and only if the rank of A is N.

(4) A is invertible if and only if there is a matrix B with $AB = I$. Then $B = A^{-1}$.

(5) A is invertible if and only if the columns of A are linearly independent.

(6) A is invertible if and only if $Ax = 0$ implies $x = 0$.

(7) A is invertible if and only if A can be transformed by elementary row operations into an upper triangular matrix having no zero entries on its main diagonal.

(8) A is invertible if and only if its determinant is not zero.

(9) A is invertible if and only if A has no zero eigenvalues.

5.3.4 Systems of Linear Equations

Consider the system of three linear equations in five unknowns given by

$$\begin{aligned}
x_1 + 2x_2 \qquad\quad + 2x_4 \quad +x_5 &= 0 \\
-x_1 - x_2 \quad +x_3 + x_4 \qquad\quad &= 0 \\
x_1 + 2x_2 \quad -3x_3 - x_4 \quad -2x_5 &= 0.
\end{aligned}$$

This system can be written in matrix form as $Ax = 0$, with A the coefficient matrix

$$A = \begin{bmatrix} 1 & 2 & 0 & 2 & 1 \\ -1 & -1 & 1 & 1 & 0 \\ 1 & 2 & -3 & -1 & -2 \end{bmatrix},$$

and $x = (x_1, x_2, x_3, x_4, x_5)^T$. Applying Gaussian elimination to this system, we obtain a second, simpler, system with the same solutions:

$$\begin{aligned} x_1 \quad\quad & - 2x_4 \quad +x_5 = 0 \\ x_2 \quad & + 2x_4 \quad\quad\quad = 0 \\ x_3 & + \ x_4 \quad +x_5 = 0. \end{aligned}$$

From this simpler system we see that the variables x_4 and x_5 can be freely chosen, with the other three variables then determined by this system of equations. The variables x_4 and x_5 are then independent, the others dependent. The variables x_1, x_2 and x_3 are then called *basic variables.* To obtain a basis of solutions we can let $x_4 = 1$ and $x_5 = 0$, obtaining the solution $x = (2, -2, -1, 1, 0)^T$, and then choose $x_4 = 0$ and $x_5 = 1$ to get the solution $x = (-1, 0, -1, 0, 1)^T$. Every solution to $Ax = 0$ is then a linear combination of these two solutions. Notice that which variables are basic and which are non-basic is somewhat arbitrary, in that we could have chosen as the non-basic variables any two whose columns are independent.

Having decided that x_4 and x_5 are the non-basic variables, we can write the original matrix A as $A = \begin{bmatrix} B & N \end{bmatrix}$, where B is the square invertible matrix

$$B = \begin{bmatrix} 1 & 2 & 0 \\ -1 & -1 & 1 \\ 1 & 2 & -3 \end{bmatrix},$$

and N is the matrix

$$N = \begin{bmatrix} 2 & 1 \\ 1 & 0 \\ -1 & -2 \end{bmatrix}.$$

With $x_B = (x_1, x_2, x_3)^T$ and $x_N = (x_4, x_5)^T$ we can write

$$Ax = Bx_B + Nx_N = 0,$$

so that

$$x_B = -B^{-1}Nx_N.$$

5.3.5 Real and Complex Systems of Linear Equations

A system $Ax = b$ of linear equations is called a *complex system*, or a *real system* if the entries of A, x and b are complex, or real, respectively.

For any matrix A, we denote by A^T and $A^\dagger$ the transpose and conjugate transpose of A, respectively.

Any complex system can be converted to a real system in the following way. A complex matrix A can be written as $A = A_1 + iA_2$, where A_1 and A_2 are real matrices and $i = \sqrt{-1}$. Similarly, $x = x^1 + ix^2$ and $b = b^1 + ib^2$, where x^1, x^2, b^1 and b^2 are real vectors. Denote by $\tilde{A}$ the real matrix

$$\tilde{A} = \begin{bmatrix} A_1 & -A_2 \\ A_2 & A_1 \end{bmatrix},$$

by $\tilde{x}$ the real vector

$$\tilde{x} = \begin{bmatrix} x^1 \\ x^2 \end{bmatrix},$$

and by $\tilde{b}$ the real vector

$$\tilde{b} = \begin{bmatrix} b^1 \\ b^2 \end{bmatrix}.$$

Then x satisfies the system $Ax = b$ if and only if $\tilde{x}$ satisfies the system $\tilde{A}\tilde{x} = \tilde{b}$.

The matrices $\tilde{A}$, $\tilde{x}$ and $\tilde{b}$ are in *block-matrix form*, meaning that the entries of these matrices are described in terms of smaller matrices. This is a convenient shorthand that we shall use repeatedly in this text. When we write $\tilde{A}\tilde{x} = \tilde{b}$, we mean

$$A_1x^1 - A_2x^2 = b^1,$$

and

$$A_2x^1 + A_1x^2 = b^2.$$

Definition 5.9 *A square matrix A is* symmetric *if $A^T = A$ and* Hermitian *if $A^\dagger = A$.*

Definition 5.10 *A nonzero vector x is said to be an* eigenvector *of the square matrix A if there is a scalar λ such that $Ax = \lambda x$. Then λ is said to be an* eigenvalue *of A.*

If x is an eigenvector of A with eigenvalue λ, then the matrix $A - \lambda I$ has no inverse, so its determinant is zero; here I is the identity matrix with ones on the main diagonal and zeros elsewhere. Solving for the roots of the determinant is one way to calculate the eigenvalues of A. For example, the eigenvalues of the Hermitian matrix

$$B = \begin{bmatrix} 1 & 2+i \\ 2-i & 1 \end{bmatrix}$$

are $\lambda = 1 + \sqrt{5}$ and $\lambda = 1 - \sqrt{5}$, with corresponding eigenvectors $u = (\sqrt{5}, 2 - i)^T$ and $v = (\sqrt{5}, i - 2)^T$, respectively. Then $\tilde{B}$ has the same eigenvalues, but both with multiplicity two. Finally, the associated eigenvectors of $\tilde{B}$ are

$$\begin{bmatrix} u^1 \\ u^2 \end{bmatrix},$$

and

$$\begin{bmatrix} -u^2 \\ u^1 \end{bmatrix},$$

for $\lambda = 1 + \sqrt{5}$, and

$$\begin{bmatrix} v^1 \\ v^2 \end{bmatrix},$$

and

$$\begin{bmatrix} -v^2 \\ v^1 \end{bmatrix},$$

for $\lambda = 1 - \sqrt{5}$.

5.4 LU and QR Factorization

Let S be a real N by N matrix. Two important methods for solving the system $Sx = b$, the LU factorization and the QR factorization, involve factoring the matrix S and thereby reducing the problem to finding the solutions of simpler systems.

In the LU factorization, we seek a lower triangular matrix L and an upper triangular matrix U so that $S = LU$. We then solve $Sx = b$ by solving $Lz = b$ and $Ux = z$.

In the QR factorization, we seek an orthogonal matrix Q, that is, $Q^T = Q^{-1}$, and an upper triangular matrix R so that $S = QR$. Then we solve $Sx = b$ by solving the upper triangular system $Rx = Q^T b$.

5.5 The LU Factorization

The matrix

$$S = \begin{bmatrix} 2 & 1 & 1 \\ 4 & 1 & 0 \\ -2 & 2 & 1 \end{bmatrix}$$

can be reduced to the upper triangular matrix

$$U = \begin{bmatrix} 2 & 1 & 1 \\ 0 & -1 & -2 \\ 0 & 0 & -4 \end{bmatrix}$$

through three elementary row operations: First, add -2 times the first row to the second row; second, add the first row to the third row; finally, add three times the new second row to the third row. Each of these row operations can be viewed as the result of multiplying on the left by the matrix obtained by applying the same row operation to the identity matrix. For example, adding -2 times the first row to the second row can be achieved by multiplying A on the left by the matrix

$$L_1 = \begin{bmatrix} 1 & 0 & 0 \\ -2 & 1 & 0 \\ 0 & 0 & 1 \end{bmatrix};$$

note that the inverse of L_1 is

$$L_1^{-1} = \begin{bmatrix} 1 & 0 & 0 \\ 2 & 1 & 0 \\ 0 & 0 & 1 \end{bmatrix}.$$

We can write

$$L_3 L_2 L_1 S = U,$$

where L_1, L_2, and L_3 are the matrix representatives of the three elementary row operations. Therefore, we have

$$S = L_1^{-1} L_2^{-1} L_3^{-1} U = LU.$$

This is the *LU factorization* of S. As we just saw, the LU factorization can be obtained along with the Gauss elimination.

5.5.1 A Shortcut

There is a shortcut we can take in calculating the LU factorization. We begin with the identity matrix I, and then, as we perform a row operation, for example, adding -2 times the first row to the second row, we put the number 2, the multiplier just used, but with a sign change, in the second row, first column, the position of the entry of S that was just converted to zero. Continuing in this fashion, we build up the matrix L as

$$L = \begin{bmatrix} 1 & 0 & 0 \\ 2 & 1 & 0 \\ -1 & -3 & 1 \end{bmatrix},$$

so that

$$S = \begin{bmatrix} 2 & 1 & 1 \\ 4 & 1 & 0 \\ -2 & 2 & 1 \end{bmatrix} = \begin{bmatrix} 1 & 0 & 0 \\ 2 & 1 & 0 \\ -1 & -3 & 1 \end{bmatrix} \begin{bmatrix} 2 & 1 & 1 \\ 0 & -1 & -2 \\ 0 & 0 & -4 \end{bmatrix}.$$

The entries of the main diagonal of L will be all ones. If we want the same to be true of U, we can rescale the rows of U and obtain the factorization $S = LDU$, where D is a diagonal matrix.

5.5.2 A Warning!

We have to be careful when we use the shortcut, as we illustrate now. For the purpose of this discussion let's use the terminology $R_i + aR_j$ to mean the row operation that adds a times the jth row to the ith row, and aR_i to mean the operation that multiplies the ith row by a. Now we transform S to an upper triangular matrix U using the row operations

(1) $\frac{1}{2}R_1$;

(2) $R_2 + (-4)R_1$;

(3) $R_3 + 2R_1$;

(4) $R_3 + 3R_2$;

(5) $(-1)R_2$; and finally,

(6) $(\frac{-1}{4})R_3$.

We end up with

$$U = \begin{bmatrix} 1 & 1/2 & 1/2 \\ 0 & 1 & 2 \\ 0 & 0 & 1 \end{bmatrix}.$$

If we use the shortcut to form the lower triangular matrix L, we find that

$$L = \begin{bmatrix} 2 & 0 & 0 \\ 4 & -1 & 0 \\ -2 & -3 & -4 \end{bmatrix}.$$

Let's go through how we formed L from the row operations listed above. We get $L_{11} = 2$ from the first row operation, $L_{21} = 4$ from the second, $L_{31} = -2$ from the third, $L_{32} = -3$ from the fourth, $L_{22} = -1$ from the fifth, and $L_{33} = \frac{-1}{4}$ from the sixth. But, if we multiple LU we do not get back S! The problem is that we performed the fourth operation, adding to the third row three times the second row, before the $(2, 2)$ entry was rescaled to one. Suppose, instead, we do the row operations in this order:

(1) $\frac{1}{2}R_1$;

(2) $R_2 + (-4)R_1$;

(3) $R_3 + 2R_1$;

(4) $(-1)R_2$;

(5) $R_3 - 3R_2$; and finally,

(6) $(\frac{-1}{4})R_3$.

Then the entry L_{32} becomes 3, instead of -3, and now $LU = S$. The message is that if we want to use the shortcut and we plan to rescale the diagonal entries of U to be one, we should rescale a given row prior to adding any multiple of that row to another row; otherwise, we can get the wrong L. The problem is that certain elementary matrices associated with row operations do not commute.

We just saw that

$$L = L_1^{-1}L_2^{-1}L_3^{-1}.$$

However, when we form the matrix L simultaneously with performing the row operations, we are, in effect, calculating

$$L_3^{-1}L_2^{-1}L_1^{-1}.$$

Most of the time the order doesn't matter, and we get the correct L anyway. But this is not always the case. For example, if we perform the operation $\frac{1}{2}R_1$, followed by $R_2 + (-4)R_1$, this is not the same as doing $R_2 + (-4)R_1$, followed by $\frac{1}{2}R_1$.

With the matrix L_1 representing the operation $\frac{1}{2}R_1$ and the matrix L_2 representing the operation $R_2 + (-4)R_1$, we find that storing a 2 in the $(1,1)$ position, and then a $+4$ in the $(1,2)$ position as we build L is not equivalent to multiplying the identity matrix by $L_2^{-1}L_1^{-1}$ but rather multiplying the identity matrix by

$$(L_1^{-1}L_2^{-1}L_1)L_1^{-1} = L_1^{-1}L_2^{-1},$$

which is the correct order.

To illustrate this point, consider the matrix S given by

$$S = \begin{bmatrix} 2 & 1 & 1 \\ 4 & 1 & 0 \\ 0 & 0 & 1 \end{bmatrix}.$$

In the first instance, we perform the row operations $R_2 + (-2)R_1$, followed by $\frac{1}{2}R_1$ to get

$$U = \begin{bmatrix} 1 & 0.5 & 0.5 \\ 0 & -1 & -2 \\ 0 & 0 & 1 \end{bmatrix}.$$

Using the shortcut, the matrix L becomes

$$L = \begin{bmatrix} 2 & 0 & 0 \\ 2 & 1 & 0 \\ 0 & 0 & 1 \end{bmatrix},$$

but we do not get $S = LU$. We do have $U = L_2 L_1 S$, where

$$L_1 = \begin{bmatrix} 1 & 0 & 0 \\ -2 & 1 & 0 \\ 0 & 0 & 1 \end{bmatrix},$$

and

$$L_2 = \begin{bmatrix} 0.5 & 0 & 0 \\ 0 & 1 & 0 \\ 0 & 0 & 1 \end{bmatrix},$$

so that $S = L_1^{-1} L_2^{-1} U$ and the correct L is

$$L = L_1^{-1} L_2^{-1} = \begin{bmatrix} 2 & 0 & 0 \\ 4 & 1 & 0 \\ 0 & 0 & 1 \end{bmatrix}.$$

But when we use the shortcut to generate L, we effectively multiply the identity matrix first by L_1^{-1} and then by L_2^{-1}, giving the matrix $L_2^{-1} L_1^{-1}$ as our candidate for L. But $L_1^{-1} L_2^{-1}$ and $L_2^{-1} L_1^{-1}$ are not the same. But why does reversing the order of the row operations work?

When we perform $\frac{1}{2} R_1$ first, and then $R_2 + (-4) R_1$ to get U, we are multiplying S first by L_2 and then by the matrix

$$E = \begin{bmatrix} 1 & 0 & 0 \\ -4 & 1 & 0 \\ 0 & 0 & 1 \end{bmatrix}.$$

The correct L is then $L = L_2^{-1} E^{-1}$.

When we use the shortcut, we are first multiplying the identity by the matrix L_2^{-1} and then by a second matrix that we shall call J; the correct L must then be $L = J L_2^{-1}$. The matrix J is not E^{-1}, but

$$J = L_2^{-1} E^{-1} L_2,$$

so that

$$L = J L_2^{-1} = L_2^{-1} E^{-1} L_2 L_2^{-1} = L_2^{-1} E^{-1},$$

which is correct.

Note that it may not be possible to obtain $A = LDU$ without first permuting the rows of A; in such cases we obtain $PA = LDU$, where P is obtained from the identity matrix by permuting rows.

Suppose that we have to solve the system of linear equations $Ax = b$. Once we have the LU factorization, it is a simple matter to find x: first, we solve the system $Lz = b$, and then solve $Ux = z$. Because both L and U are triangular, solving these systems is a simple matter. Obtaining the LU factorization is often better than finding A^{-1}; when A is banded, that is, has nonzero values only for the main diagonal and a few diagonals on either side, the L and U retain that banded property, while A^{-1} does not.

If A is real and symmetric, and if $A = LDU$, then $U = L^T$, so we have $A = LDL^T$. If, in addition, the nonzero entries of D are positive, then we can write

$$A = (L\sqrt{D})(L\sqrt{D})^T,$$

which is the Cholesky Decomposition of A.

5.5.3 The QR Factorization and Least Squares

The least-squares solution of $Ax = b$ is the solution of $A^TAx = A^Tb$. Once we have $A = QR$, we have $A^TA = R^TQ^TQR = R^TR$, so we find the least-squares solution easily, by solving $R^Tz = A^Tb$, and then $Rx = z$. Note that $A^TA = R^TR$ is the Cholesky decomposition of A^TA.

5.6 Exercises

Ex. 5.1 *Prove that the zero vector of a vector space is unique, that each v has a unique additive inverse $-v$, and that $(-1)v = -v$, for all v.*

Ex. 5.2 *Let $W = \{w^1, ..., w^N\}$ be a spanning set for a subspace $\mathcal{S}$ in $\mathbb{R}^I$, and $V = \{v^1, ..., v^M\}$ a linearly independent subset of $\mathcal{S}$. Let A be the matrix whose columns are the v^m, B the matrix whose columns are the w^n. Show that there is an N by M matrix C such that $A = BC$. Prove Lemma 5.1 by showing that, if $M > N$, then there is a nonzero vector x with $Cx = Ax = 0$.*

Ex. 5.3 *Prove Proposition 5.3.*

Ex. 5.4 *Prove that, if L is invertible and lower triangular, then so is L^{-1}.*

Ex. 5.5 *Show that the symmetric matrix*

$$H = \begin{bmatrix} 0 & 1 \\ 1 & 0 \end{bmatrix}$$

cannot be written as $H = LDL^T$.

Ex. 5.6 *Show that the symmetric matrix*

$$H = \begin{bmatrix} 0 & 1 \\ 1 & 0 \end{bmatrix}$$

cannot be written as $H = LU$, *where* L *is lower triangular,* U *is upper triangular, and both are invertible.*

Ex. 5.7 *Let* F *be an invertible matrix that is the identity matrix, except for column* s. *Show that* $E = F^{-1}$ *is also the identity matrix, except for the entries in column* s, *which can be explicitly calculated from those of* F.

Chapter 6

Linear Programming

6.1 Chapter Summary

The term *linear programming* (LP) refers to the problem of optimizing a linear function of several variables over linear equality or inequality constraints. In this chapter we present the problem and discuss weak and strong duality and the simplex method. For a much more detailed treatment of linear programming, consult [164].

6.2 Primal and Dual Problems

The fundamental problem in linear programming is to minimize the function

$$f(x) = \langle c, x \rangle = c \cdot x = c^T x = z,$$

over the *feasible set* F, that is, the convex set of all $x \geq 0$ with $Ax = b$. This is the *primal problem* in *standard form*, denoted PS. The set F is then the feasible set for PS, and any x in F is called a feasible solution for PS, or said to be feasible for PS. We shall use theorems of the alternative to establish the basic facts about LP problems.

Shortly, we shall present an algebraic description of the extreme points of the feasible set F, in terms of *basic feasible solutions*, show that there are at most finitely many extreme points of F and that every member of F can be written as a convex combination of the extreme points, plus a direction of unboundedness. These results are also used to prove the basic theorems about linear programming problems and to describe the simplex algorithm.

Associated with the basic problem in LP, called the *primary problem*, there is a second problem, the *dual problem*. Both of these problems can be written in two equivalent ways, the canonical form and the standard form.

6.2.1 An Example

Consider the problem of maximizing the function $f(x_1 x_2) = x_1 + 2x_2$, over all $x_1 \geq 0$ and $x_2 \geq 0$, for which the inequalities

$$x_1 + x_2 \leq 40,$$

and

$$2x_1 + x_2 \leq 60$$

are satisfied. The set of points satisfying all four inequalities is the region of $\mathbb{R}^2$ bounded by the quadrilateral with vertices $(0, 0)$, $(30, 0)$, $(20, 20)$, and $(0, 40)$; draw a picture. Since the level curves of the function f are straight lines, the maximum value must occur at one of these vertices; in fact, it occurs at $(0, 40)$ and the maximum value of f over the constraint set is 80. Rewriting the problem as minimizing the function $-x_1 - 2x_2$, subject to $x_1 \geq 0$, $x_2 \geq 0$,

$$-x_1 - x_2 \geq -40,$$

and

$$-2x_1 - x_2 \geq -60,$$

the problem is now in what is called *primal canonical form*.

6.2.2 Canonical and Standard Forms

Let b and c be fixed vectors, A a fixed I by J matrix, and x a vector variable in $\mathbb{R}^J$. The problem

$$\text{minimize}\, z = c^T x, \,\text{subject to}\, Ax \geq b, \, x \geq 0 \ \ \text{(PC)}$$

is the so-called *primary problem* of LP, in *canonical form*. The *dual problem* in canonical form is

$$\text{maximize}\, w = b^T y, \,\text{subject to}\, A^T y \leq c, \, y \geq 0. \ \ \text{(DC)}$$

The primary problem, in *standard form*, is

$$\text{minimize}\, z = c^T x, \,\text{subject to}\, Ax = b, \, x \geq 0 \ \ \text{(PS)}$$

with the dual problem in standard form given by

$$\text{maximize}\, w = b^T y, \,\text{subject to}\, A^T y \leq c. \ \ \text{(DS)}$$

Notice that the dual problem in standard form does not require that y be nonnegative. Note also that PS makes sense only if the system $Ax = b$ has solutions. For that reason, we shall assume, for the standard problems, that the I by J matrix A has at least as many columns as rows, so $J \geq I$, and A has full rank I.

The primal problem PC can be rewritten in dual canonical form, as

$$\text{maximize}\, (-c)^T x, \,\text{subject to}\, (-A)x \leq -b, \, x \geq 0.$$

The corresponding primal problem is then

$$\text{minimize}\, (-b)^T y, \,\text{subject to}\, (-A)^T y \geq -c, \, y \geq 0,$$

which can obviously be rewritten as problem DC. This "symmetry"of the canonical forms will be useful later in proving strong duality theorems.

6.2.3 From Canonical to Standard and Back

If we are given the primary problem in canonical form, we can convert it to standard form by augmenting the variables, that is, by introducing the *slack variables*

$$u_i = (Ax)_i - b_i,$$

for $i = 1, ..., I$, and rewriting $Ax \geq b$ as

$$\tilde{A}\tilde{x} = b,$$

for $\tilde{A} = [A \quad -I]$ and $\tilde{x} = [x^T \ u^T]^T$. If PC has a feasible solution, then so does its PS version. If the corresponding dual problem DC is feasible, then so is its DS version; the new c is $\tilde{c} = [c^T \ 0^T]^T$. The quantities z and w remain unchanged.

If we are given the primary problem in standard form, we can convert it to canonical form by writing the equations as inequalities, that is, by replacing $Ax = b$ with the two matrix inequalities $Ax \geq b$, and $(-A)x \geq -b$ and writing $\tilde{A}x \geq \tilde{b}$, where $\tilde{A} = [A^T \ \ -A^T]^T$ and $\tilde{b} = [b^T \ \ -b^T]^T$. If the problem PS is feasible, then so is its PC version. If the corresponding dual problem DS is feasible, so is DC, where now the new y is $\tilde{y} = [u^T \ \ -v^T]^T$, where $u_i = \max\{y_i, 0\}$ and $v_i = y_i - u_i$. Again, the z and w remain unchanged.

6.3 Converting a Problem to PS Form

The following example, taken from [137], illustrates the modifications we may need to make to convert a linear-programming problem to PS form.

Suppose that we want to maximize the function

$$f(x_1, x_2, x_3, x_4) = 5x_1 + 3x_2 + 3x_3 + x_4,$$

subject to

$$2x_2 + x_4 = 2,$$

$$x_1 + x_2 + x_4 \leq 3,$$

$$-x_1 - 2x_2 + x_3 \geq 1,$$

and

$$x_1 \leq 0, x_2, x_3 \geq 0.$$

First, we introduce stack variables x_5 and x_7, and surplus variable x_6 and write

$$2x_2 + x_4 = 2,$$

$$x_1 + x_2 + x_4 + x_5 = 3,$$

$$-x_1 - 2x_2 + x_3 - x_6 = 1,$$

and

$$x_1 + x_7 = 0.$$

Then we also have

$$x_2, x_3, x_5, x_6, x_7 \geq 0.$$

Because the variables x_1 and x_4 are not bounded below, we replace x_1 with

$-x_7$ and x_4 with $2 - 2x_2$ throughout. Now the problem is to minimize the function

$$z = x_2 + 3x_3 - 5x_7,$$

subject to

$$-x_2 + x_5 - x_7 = 1,$$

$$-2x_2 + x_3 - x_6 + x_7 = 1,$$

and

$$x_2, x_3, x_5, x_6, x_7 \geq 0.$$

This is the problem in PS form.

6.4 Duality Theorems

The main topics in linear programming are the duality theorems and the simplex algorithm. In this section we consider duality theorems.

6.4.1 Weak Duality

Consider the problems PS and DS. Say that x is *feasible* for PS if $x \geq 0$ and $Ax = b$. Let F be the set of such feasible x. Say that y is *feasible* for DS if $A^T y \leq c$. When it is clear from the context which problems we are discussing, we shall simply say that x and y are feasible.

The *Weak Duality Theorem* is the following:

Theorem 6.1 *Let x and y be feasible vectors. Then*

$$z = c^T x \geq b^T y = w.$$

Corollary 6.1 *If z is not bounded below, then there are no feasible y.*

Corollary 6.2 *If x and y are both feasible, and $z = w$, then both x and y are optimal for their respective problems.*

The proof of the theorem and its corollaries are left as exercises.

6.4.2 Primal-Dual Methods

The nonnegative quantity $c^Tx - b^Ty$ is called the *duality gap* for x and y. The *complementary slackness condition* says that, for optimal x and y, we have

$$x_j(c_j - (A^Ty)_j) = 0,$$

for each j. Introducing the *slack variables* $s_j \geq 0$, for $j = 1, ..., J$, we can write the dual problem constraint $A^Ty \leq c$ as $A^Ty + s = c$. Then the complementary slackness conditions $x_js_j = 0$ for each j are equivalent to $z = w$, so the duality gap is zero. Primal-dual algorithms for solving linear programming problems are based on finding sequences of vectors $\{x^k\}$, $\{y^k\}$, and $\{s^k\}$ that drive $x_j^k s_j^k$ down to zero, and therefore, the duality gaps down to zero [164].

6.4.3 Strong Duality

The *Strong Duality Theorems* make a stronger statement. One such theorem is the following.

Theorem 6.2 *If one of the problems PS or DS has an optimal solution, then so does the other and $z = w$ for the optimal vectors.*

Another strong duality theorem is due to David Gale [115].

Theorem 6.3 (Gale's Strong Duality Theorem) *If both problems PC and DC have feasible solutions, then both have optimal solutions and the optimal values are equal.*

6.5 A Basic Strong Duality Theorem

In this section we state and prove a basic strong duality theorem that has, as corollaries, both Theorem 6.2 and Gale's Strong Duality Theorem 6.3, as well as other theorems of this type. The proof of this basic strong duality theorem is an immediate consequence of Farkas' Lemma, which we repeat here for convenience.

Theorem 6.4 (Farkas' Lemma)[110] *Precisely one of the following is true:*

(1) there is $x \geq 0$ such that $Ax = b$;

(2) there is y such that $A^Ty \geq 0$ and $b^Ty < 0$.

We begin with a few items of notation. Let p be the infimum of the values $c^T x$, over all $x \geq 0$ such that $Ax = b$, with $p = \infty$ if there are no such x. Let p^* be the supremum of the values $b^T y$, over all y such that $A^T y \leq c$, with $p^* = -\infty$ if there are no such y. Let v be the infimum of the values $c^T x$, over all $x \geq 0$ such that $Ax \geq b$, with $v = \infty$ if there are no such x. Let v^* be the supremum of the values $b^T y$, over all $y \geq 0$ such that $A^T y \leq c$, with $v^* = -\infty$ if there are no such y. Our basic strong duality theorem is the following.

Theorem 6.5 (A Basic Strong Duality Theorem) *If p^* is finite, then the primal problem PS has an optimal solution $\hat{x}$ and $c^T \hat{x} = p^*$.*

Proof: Consider the system of inequalities given in block-matrix form by

$$\begin{bmatrix} -A^T & c \\ 0^T & 1 \end{bmatrix} \begin{bmatrix} r \\ \alpha \end{bmatrix} \geq \begin{bmatrix} 0 \\ 0 \end{bmatrix},$$

and

$$\begin{bmatrix} -b^T & p^* \end{bmatrix} \begin{bmatrix} r \\ \alpha \end{bmatrix} < 0.$$

Here r is a column vector and α is a real number. We show that this system has no solution.

If there is a solution with $\alpha > 0$, then $y = \frac{1}{\alpha} r$ is feasible for the dual problem DS, but $b^T y > p^*$, contradicting the definition of p^*.

If there is a solution with $\alpha = 0$, then $A^T r \leq 0$, and $b^T r > 0$. We know that the problem DS has feasible vectors, so let $\hat{y}$ be one such. Then the vectors $\hat{y} + nr$ are feasible vectors, for $n = 1, 2, ...$ But $b^T(\hat{y} + nr) \to +\infty$, as n increases, contradicting the assumption that p^* is finite.

Now, by Farkas' Lemma, there must be $\hat{x} \geq 0$ and $\beta \geq 0$ such that $A\hat{x} = b$ and $c^T \hat{x} = p^* - \beta \leq p^*$. It follows that $\hat{x}$ is optimal for the primal problem PS and $c^T \hat{x} = p^*$. ∎

Now we reap the harvest of corollaries of this basic strong duality theorem. First, recall that LP problems in standard form can be reformulated as LP problems in canonical form, and vice versa. Also recall the "symmetry" of the canonical forms; the problem PC can be rewritten in form of a DC problem, whose corresponding primal problem in canonical form is equivalent to the original DC problem. As a result, we have the following corollaries of Theorem 6.5.

Corollary 6.3 *Let p be finite. Then DS has an optimal solution $\hat{y}$ and $b^T \hat{y} = p$.*

Corollary 6.4 *Let v be finite. Then DC has an optimal solution $\hat{y}$ and $b^T \hat{y} = v$.*

Corollary 6.5 *Let v^* be finite. Then PC has an optimal solution $\hat{x}$ and $c^T\hat{x} = v^*$.*

Corollary 6.6 *Let p or p^* be finite. Then both PS and DS have optimal solutions $\hat{x}$ and $\hat{y}$, respectively, with $c^T\hat{x} = b^T\hat{y}$.*

Corollary 6.7 *Let v or v^* be finite. Then both PC and DC have optimal solutions $\hat{x}$ and $\hat{y}$, respectively, with $c^T\hat{x} = b^T\hat{y}$.*

In addition, Theorem 6.2 follows as a corollary, since if either PS or DS has an optimal solution, then one of p or p^* must be finite. Gale's Strong Duality Theorem 6.3 is also a consequence of Theorem 6.5, since, if both PC and DC are feasible, then both v and v^* must be finite.

6.6 Another Proof

We know that Theorem 6.2 is a consequence of Theorem 6.5, which, in turn, follows from Farkas' Lemma. However, it is instructive to consider an alternative proof of Theorem 6.2. For that, we need some definitions and notation.

Definition 6.1 *A point x in F is said to be a* basic feasible solution *if the columns of A corresponding to positive entries of x are linearly independent.*

Recall that, for PS, we assume that $J \geq I$ and the rank of A is I. Consequently, if, for some nonnegative vector x, the columns j for which x_j is positive are linearly independent, then x_j is positive for at most I values of j. Therefore, a basic feasible solution can have at most I positive entries. For a given set of entries, there can be at most one basic feasible solution for which precisely those entries are positive. Therefore, there can be only finitely many basic feasible solutions.

Now let x be an arbitrary basic feasible solution. Denote by B an invertible matrix obtained from A by deleting $J - I$ columns associated with zero entries of x. Note that, if x has fewer than I positive entries, then some of the columns of A associated with zero values of x_j are retained. The entries of an arbitrary vector y corresponding to the columns not deleted are called the *basic variables*. Then, assuming that the columns of B are the first I columns of A, we write $y^T = (y_B^T, y_N^T)$, and

$$A = \begin{bmatrix} B & N \end{bmatrix},$$

so that $Ay = By_B + Ny_N$, $Ax = Bx_B = b$, and $x_B = B^{-1}b$.

The following theorems are taken from the book by Nash and Sofer [164]. We begin with a characterization of the extreme points of F (recall Definition 4.22).

Theorem 6.6 *A point x is in Ext(F) if and only if x is a basic feasible solution.*

Proof: Suppose that x is a basic feasible solution, and we write $x^T = (x_B^T, 0^T)$, $A = \begin{bmatrix} B & N \end{bmatrix}$. If x is not an extreme point of F, then there are $y \neq x$ and $z \neq x$ in F, and α in $(0,1)$, with

$$x = (1-\alpha)y + \alpha z.$$

Then $y^T = (y_B^T, y_N^T)$, $z^T = (z_B^T, z_N^T)$, and $y_N \geq 0$, $z_N \geq 0$. From

$$0 = x_N = (1-\alpha)y_N + (\alpha)z_N$$

it follows that

$$y_N = z_N = 0,$$

and $b = By_B = Bz_B = Bx_B$. But, since B is invertible, we have $x_B = y_B = z_B$. This is a contradiction, so x must be in Ext(F).

Conversely, suppose that x is in Ext(F). Since x is in F, we know that $Ax = b$ and $x \geq 0$. By reordering the variables if necessary, we may assume that $x^T = (x_B^T, x_N^T)$, with $x_B > 0$ and $x_N = 0$; we do not know that x_B is a vector of length I, however, so when we write $A = \begin{bmatrix} B & N \end{bmatrix}$, we do not know that B is square.

If the columns of B are linearly independent, then, by definition, x is a basic feasible solution. If the columns of B were not linearly independent, we could construct $y \neq x$ and $z \neq x$ in F, such that

$$x = \frac{1}{2}y + \frac{1}{2}z,$$

as we now show. If $\{B_1, B_2, ..., B_K\}$ are the columns of B and are linearly dependent, then there are constants $p_1, p_2, ..., p_K$, not all zero, with

$$p_1B_1 + ... + p_KB_K = 0.$$

With $p^T = (p_1, ..., p_K)$, we have

$$B(x_B + \alpha p) = B(x_B - \alpha p) = Bx_B = b,$$

for all $\alpha \in (0,1)$. We then select α so small that both $x_B + \alpha p > 0$ and $x_B - \alpha p > 0$. Let

$$y^T = (x_B^T + \alpha p^T, 0^T)$$

and

$$z^T = (x_B^T - \alpha p^T, 0^T).$$

Therefore x is not an extreme point of F, which is a contradiction. This completes the proof. ∎

Corollary 6.8 *There are at most finitely many basic feasible solutions, so there are at most finitely many members of Ext(F).*

Theorem 6.7 *If F is not empty, then Ext(F) is not empty. In that case, let $\{v^1, ..., v^M\}$ be the members of Ext(F). Every x in F can be written as*

$$x = d + \alpha_1 v^1 + ... + \alpha_M v^M, \tag{6.1}$$

for some $\alpha_m \geq 0$, with $\sum_{m=1}^{M} \alpha_m = 1$, and some direction of unboundedness, d.

Proof: We consider only the case in which F is bounded, so there is no direction of unboundedness; the unbounded case is similar. Let x be a feasible point. If x is an extreme point, fine. If not, then x is not a basic feasible solution and the columns of A that correspond to the positive entries of x are not linearly independent. Then we can find a vector p such that $Ap = 0$ and $p_j = 0$ if $x_j = 0$. If $|\epsilon|$ is small enough, $x + \epsilon p$ is in F and $(x + \epsilon p)_j = 0$ if $x_j = 0$. Our objective now is to find another member of F that has fewer positive entries than x has.

We can alter ϵ in such a way that eventually $y = x + \epsilon p$ has at least one more zero entry than x has. To see this, let

$$-\epsilon = \frac{x_k}{p_k} = \min\Big(\frac{x_j}{p_j} | x_j > 0, p_j > 0\Big).$$

Then the vector $x + \epsilon p$ is in F and has fewer positive entries than x has. Repeating this process, we must eventually reach the point at which there is no such vector p. At this point, we have obtained a basic feasible solution, which must then be an extreme point of F. Therefore, the set of extreme points of F is not empty.

The set G of all x in F that can be written as in Equation (6.1) is a closed set. Consequently, if there is x in F that cannot be written in this way, there is a ball of radius r, centered at x, having no intersection with G. We can then repeat the previous construction to obtain a basic feasible solution that lies within this ball. But such a vector would be an extreme point of F, and so would have to be a member of G, which would be a contradiction. Therefore, every member of F can be written according to Equation (6.1). ∎

Proof of Theorem 6.2: Suppose now that x_* is a solution of the problem PS and $z_* = c^T x_*$. Without loss of generality, we may assume that x_* is a basic feasible solution, hence an extreme point of F. (Why?) Then we can write

$$x_*^T = ((B^{-1}b)^T, 0^T),$$
$$c^T = (c_B^T, c_N^T),$$

and $A = \begin{bmatrix} B & N \end{bmatrix}$. We shall show that

$$y_* = (B^{-1})^T c_B,$$

which depends on x_* via the matrix B, and

$$z_* = c^T x_* = y_*^T b = w_*.$$

Every feasible solution has the form

$$x^T = ((B^{-1}b)^T, 0^T) + ((B^{-1}Nv)^T, v^T),$$

for some $v \geq 0$. From $c^T x \geq c^T x_*$ we find that

$$(c_N^T - c_B^T B^{-1} N)(v) \geq 0,$$

for all $v \geq 0$. It follows that

$$c_N^T - c_B^T B^{-1} N = 0.$$

Now let $y_* = (B^{-1})^T c_B$, or $y_*^T = c_B^T B^{-1}$. We show that y_* is feasible for DS; that is, we show that

$$A^T y_* \leq c^T.$$

Since

$$y_*^T A = (y_*^T B, y_*^T N) = (c_B^T, y_*^T N) = (c_B^T, c_B^T B^{-1} N)$$

and

$$c_N^T \geq c_B^T B^{-1} N,$$

we have

$$y_*^T A \leq c^T,$$

so y_* is feasible for DS. Finally, we show that

$$c^T x_* = y_*^T b.$$

We have

$$y_*^T b = c_B^T B^{-1} b = c^T x_*.$$

This completes the proof. ∎

6.7 Proof of Gale's Strong Duality Theorem

As we have seen, Gale's Strong Duality Theorem 6.3 is a consequence of Theorem 6.5, and so follows from Farkas' Lemma. Gale's own proof, which we give below, is somewhat different, in that he uses Farkas' Lemma to obtain Theorem 4.11, and then the results of Theorem 4.11 to prove Theorem 6.3.

We show that there are nonnegative vectors x and y such that $Ax \geq b$, $A^T y \leq c$, and $b^T y - c^T x \geq 0$. It will then follow that $z = c^T x = b^T y = w$, so that x and y are both optimal. In matrix notation, we want to find $x \geq 0$ and $y \geq 0$ such that

$$\begin{bmatrix} A & 0 \\ 0 & -A^T \\ -c^T & b^T \end{bmatrix} \begin{bmatrix} x \\ y \end{bmatrix} \geq \begin{bmatrix} b \\ -c \\ 0 \end{bmatrix}. \tag{6.2}$$

In order to use Theorem 4.11, we rewrite (6.2) as

$$\begin{bmatrix} -A & 0 \\ 0 & A^T \\ c^T & -b^T \end{bmatrix} \begin{bmatrix} x \\ y \end{bmatrix} \leq \begin{bmatrix} -b \\ c \\ 0 \end{bmatrix}. \tag{6.3}$$

We assume that there are no $x \geq 0$ and $y \geq 0$ for which the inequalities in (6.3) hold. Then, according to Theorem 4.11, there are nonnegative vectors s and t, and nonnegative scalar ρ such that

$$\begin{bmatrix} -A^T & 0 & c \\ 0 & A & -b \end{bmatrix} \begin{bmatrix} s \\ t \\ \rho \end{bmatrix} \geq 0,$$

and

$$\begin{bmatrix} -b^T & c^T & 0 \end{bmatrix} \begin{bmatrix} s \\ t \\ \rho \end{bmatrix} < 0.$$

Note that ρ cannot be zero, for then we would have $A^T s \leq 0$ and $At \geq 0$. Taking feasible vectors x and y, we would find that $s^T Ax \leq 0$, which implies that $b^T s \leq 0$, and $t^T A^T y \geq 0$, which implies that $c^T t \geq 0$. Therefore, we could not also have $c^T t - b^T s < 0$.

Writing out the inequalities, we have

$$\rho c^T t \geq s^T At \geq s^T(\rho b) = \rho s^T b.$$

Using $\rho > 0$, we find that

$$c^T t \geq b^T s,$$

which is a contradiction. Therefore, there do exist $x \geq 0$ and $y \geq 0$ such that $Ax \geq b$, $A^T y \leq c$, and $b^T y - c^T x \geq 0$. ∎

6.8 Some Examples

We give two well known examples of LP problems.

6.8.1 The Diet Problem

There are nutrients indexed by $i = 1, ..., I$ and our diet must contain at least b_i units of the ith nutrient. There are J foods, indexed by $j = 1, ..., J$, and one unit of the jth food costs c_j dollars and contains A_{ij} units of the ith nutrient. The problem is to minimize the cost, while obtaining at least the minimum amount of each nutrient.

Let $x_j \geq 0$ be the amount of the jth food that we consume. Then we need $Ax \geq b$, where A is the matrix with entries A_{ij}, b is the vector with entries b_i and x is the vector with entries $x_j \geq 0$. With c the vector with entries c_j, the total cost of our food is $z = c^T x$. The problem is then to minimize $z = c^T x$, subject to $Ax \geq b$ and $x \geq 0$. This is the primary LP problem, in canonical form.

6.8.2 The Transport Problem

We must ship products from sources to destinations. There are I sources, indexed by $i = 1, ..., I$, and J destinations, indexed by $j = 1, ..., J$. There are a_i units of product at the ith source, and we must have at least b_j units reaching the jth destination. The customer will pay C_{ij} dollars to get one unit from i to j. Let x_{ij} be the number of units of product to go from the ith source to the jth destination. The producer wishes to maximize income, that is,

$$\text{maximize} \sum_{i,j} C_{ij} x_{ij},$$

subject to

$$x_{ij} \geq 0,$$

$$\sum_{i=1}^{I} x_{ij} \geq b_j,$$

and

$$\sum_{j=1}^{J} x_{ij} \leq a_i.$$

Obviously, we must assume that

$$\sum_{i=1}^{I} a_i \geq \sum_{j=1}^{J} b_j.$$

This problem is not yet in the form of the LP problems considered so far. It also introduces a new feature, namely, it may be necessary to have x_{ij} a nonnegative integer, if the products exist only in whole units. This leads to *integer programming.*

6.9 The Simplex Method

In this section we sketch the main ideas of the simplex method. For further details see [164].

Begin with $\hat{x}$, a basic feasible solution of PS. Assume, as previously, that

$$A = \begin{bmatrix} B & N \end{bmatrix},$$

where B is an I by I invertible matrix obtained by deleting from A some (but perhaps not all) columns associated with zero entries of $\hat{x}$. As before, we assume the variables have been ordered so that the zero entries of $\hat{x}$ have the highest index values. The entries of an arbitrary x corresponding to the first I columns are the basic variables. We write $x^T = (x_B^T, x_N^T)$, so that $\hat{x}_N = 0$, $A\hat{x} = B\hat{x}_B = b$, and $\hat{x}_B = B^{-1}b$. The current value of z is

$$\hat{z} = c_B^T \hat{x}_B = c_B^T B^{-1} b.$$

We are interested in what happens to z as x_N takes on positive entries.

For any feasible x we have $Ax = b = Bx_B + Nx_n$, so that

$$x_B = B^{-1}b - B^{-1}Nx_N,$$

and

$$z = c^T x = c_B^T x_B + c_N^T x_N = c_B^T(B^{-1}b - B^{-1}Nx_N) + c_N^T x_N.$$

Therefore,

$$z = c_B^T B^{-1} b + (c_N^T - c_B^T B^{-1} N) x_N = \hat{z} + r^T x_N,$$

where

$$r^T = (c_N^T - c_B^T B^{-1} N).$$

The vector r is called the *reduced cost vector*. We define the vector $y^T = c_B^T B^{-1}$ of *simplex multipliers*, and write

$$z - \hat{z} = r^T x_N = (c_N^T - y^T N)x_N.$$

We are interested in how z changes as we move away from $\hat{x}$ and permit x_N to have positive entries.

If x_N is nonzero, then z changes by $r^T x_N$. Therefore, if $r \geq 0$, the current $\hat{z}$ cannot be made smaller by letting x_N have some positive entries; the current $\hat{x}$ is then optimal. Initially, at least, r will have some negative entries, and we use these as a guide in deciding how to select x_N.

Keep in mind that the vectors x_N and r have length $J - I$ and the jth column of N is the $(I + j)$th column of A.

Select an index j such that

$$r_j < 0,$$

and r_j is the most negative of the negative entries of r. Then x_{I+j} is called the *entering variable*. Compute $d^j = B^{-1}a^j$, where a^j is the $(I + j)$th column of A, which is the jth column of N. If we allow $(x_N)_j = x_{I+j}$ to be positive, then

$$x_B = B^{-1}b - x_{I+j}B^{-1}a^j = B^{-1}b - x_{I+j}d^j.$$

We need to make sure that x_B remains nonnegative, so we need

$$(B^{-1}b)_i - x_{I+j}d_i^j \geq 0,$$

for all indices $i = 1, ..., I$. If the ith entry d_i^j is negative, then $(x_B)_i$ increases as x_{I+j} becomes positive; if $d_i^j = 0$, then $(x_B)_i$ remains unchanged. The problem arises when d_i^j is positive.

Find an index s in $\{1, ..., I\}$ for which

$$\frac{(B^{-1}b)_s}{d_s^j} = \min\left\{\frac{(B^{-1}b)_i}{d_i^j} : d_i^j > 0\right\}. \tag{6.4}$$

Then x_s is the *leaving variable*, replacing x_{I+j}; that is, the new set of indices corresponding to new basic variables will now include $I + j$, and no longer include s. The new entries of $\hat{x}$ are $\hat{x}_s = 0$ and

$$\hat{x}_{I+j} = \frac{(B^{-1}b)_s}{d_s^j}.$$

We then rearrange the columns of A to redefine B and N, and rearrange the positions of the entries of x, to get the new basic variables vector x_B, the new x_N and the new c. Then we repeat the process.

6.10 Yet Another Proof

Previously, we showed that Theorem 6.2 can be obtained as a corollary of the Basic Strong Duality Theorem 6.5. Now we show that Theorem 6.2 follows from the calculations involved in the simplex method.

Suppose that the problem PS has an optimal solution, $\hat{x}$. Then the vector $\hat{y}$ with $\hat{y}^T = c_B^T B^{-1}$ becomes a feasible vector for the dual problem and an optimal solution for DS. Clearly, since $\hat{x}_B = B^{-1}b$, we have

$$\hat{z} = c^T\hat{x} = c_B^T B^{-1} b = \hat{y}^T b.$$

We know that

$$r^T = c_N^T - \hat{y}^T N \geq 0,$$

so that

$$\hat{y}^T A = [\hat{y}^T B \; \hat{y}^T N] \leq [c_B^T \; c_N^T] = c^T.$$

Therefore, $\hat{y}$ is feasible for DS and

$$\hat{w} = \hat{y}^T b = \hat{z}.$$

It is interesting to note that this proof does not rely on the Bolzano–Weierstrass Theorem or any other equivalent statement of compactness. The Basic Strong Duality Theorem 6.5 relies on Farkas' Lemma, which uses orthogonal projection, and therefore the Bolzano–Weierstrass Theorem. We could say, therefore, that Theorem 6.2 is a relatively "weak" strong duality theorem.

6.11 The Sherman–Morrison–Woodbury Identity

It is helpful to note that when the columns of A are rearranged and a new B is defined, the new B differs from the old B in only one column. Therefore

$$B_{\text{new}} = B_{\text{old}} - uv^T, \tag{6.5}$$

where u is the column vector that equals the old column minus the new one, and v is the column of the identity matrix corresponding to the column of B_{old} being altered. In Exercise 6.5 the reader is asked to prove that

$$1 - v^T B_{\text{old}}^{-1} u \neq 0.$$

Once we know that, the inverse of B_{new} can be obtained fairly easily from the inverse of B_{old} using the Sherman–Morrison–Woodbury Identity.

In Exercise 6.4 the reader is asked to show that, if B is invertible, then

$$(B - uv^T)^{-1} = B^{-1} + \alpha^{-1}(B^{-1}u)(v^T B^{-1}),$$

whenever

$$\alpha = 1 - v^T B^{-1} u \neq 0,$$

and, if $\alpha = 0$, then the matrix $B - uv^T$ has no inverse. We shall illustrate this in the example below.

For large-scale problems, issues of storage, computational efficiency and numerical accuracy become increasingly important [203]. For such problems, other ways of updating the matrix B^{-1} are used.

Let F be the identity matrix, except for having the vector d^j as column s. It is easy to see that $B^{\text{new}} = BF$, so that $(B^{\text{new}})^{-1} = EB^{-1}$, where $E = F^{-1}$. In Exercise 5.7 you are asked to show that E is also the identity matrix, except for the entries in column s, which can be explicitly calculated (see [164]). Therefore, as the simplex iteration proceeds, the next $(B^{\text{new}})^{-1}$ can be represented as

$$(B^{\text{new}})^{-1} = E_k E_{k-1} \cdots E_1 B^{-1},$$

where B is the original matrix selected at the beginning of the calculations, and the other factors are the E matrices used at each step.

Another approach is to employ the LU-decomposition method for solving systems of linear equations, with numerically stable procedures for updating the matrices L and U as the columns of B are swapped. Finding methods for doing this is an active area of research [203].

6.12 An Example of the Simplex Method

Consider once again the problem of maximizing the function $f(x_1, x_2) = x_2 + 2x_2$, over all $x_1 \geq 0$ and $x_2 \geq 0$, for which the inequalities

$$x_1 + x - 2 \leq 40,$$

and

$$2x_1 + x - 2 \leq 60$$

are satisfied. As we saw previously, this problem, in primary canonical form, is the following: Minimize the function $-x_1 - 2x_2$, subject to $x_1 \geq 0$, $x_2 \geq 0$,

$$-x_1 - x_2 \geq -40,$$

and

$$-2x_1 - x - 2 \geq -60.$$

With $b^T = (-40, -60)^T$, and, temporarily, $c^T = (1, 2)^T$, A the matrix

$$A = \begin{bmatrix} -1 & -1 \\ -2 & -1 \end{bmatrix},$$

and the vector $x = (x_1, x_2)^T$, the function to be minimized is $z = c^T x$ and the constraints are $Ax \geq b$, and the vector x has nonnegative entries, that is, $x \geq 0$. This is the problem in PC form.

We say that the definitions of c^T, A and $x = (x_1, y_2)^T$ are only temporary because, first of all, once we have converted the problem to PS form, the vector x will change size, and have more entries, so A will then have more columns and c^T more entries, and second, at each step of the simplex algorithm, two of the entries of x change places, so two columns of A must be switched and two entries of c^T must also be exchanged.

Before applying the simplex algorithm, we convert this problem to PS form using slack variables. In PS form, the problem is to minimize the function $-x_1 - 2x_2$, subject to

$$-x_1 - x_2 - x_3 = -40,$$

$$-2x_1 - x_2 - x - 4 = -60,$$

and

$$x_1, x_2, x_3, x_4 \geq 0.$$

The four vertices of the feasible region can be described using these four variables as follows: the points $(0, 0)$, $(30, 0)$, $(20, 20)$, and $(0, 40)$ in the x_1, x_2 plane become $(0, 0, 40, 60)$, $(30, 0, 10, 0)$, $(20, 20, 0, 0)$ and $(0, 40, 0, 20)$, respectively, and these four points in the variables x_1, x_2, x_3, and x_4 are the basic feasible solutions. Note that each one has no more than $I = 2$ positive entries.

We begin the simplex algorithm by selecting one of the four basic feasible solutions; here we select the point $(20, 20, 0, 0)$. The variables x_1 and x_2 are now the basic variables, since they correspond to positive entries of our starting vertex. Since x_1 and x_2 were already the first two entries of the vector x, there is no need to redefine the variables in this case.

The constraint $Ax \geq b$ is now $Ax = b$, where the matrix A has become

$$A = \begin{bmatrix} -1 & -1 & -1 & 0 \\ -2 & -1 & 0 & -1 \end{bmatrix}.$$

Since x_1 and x_2 are the basic variables, the matrix B is

$$B = \begin{bmatrix} -1 & -1 \\ -2 & -1 \end{bmatrix},$$

with inverse

$$B^{-1} = \begin{bmatrix} 1 & -1 \\ -2 & 1 \end{bmatrix},$$

and the matrix N is

$$N = \begin{bmatrix} -1 & 0 \\ 0 & -1 \end{bmatrix}.$$

The vector b is $b = (-40, -60)^T$. For a general vector $x = (x_1, x_2, x_3, x_4)^T$, we write $x_B = (x_1, x_2)^T$ and $x_N = (x_3, x_4)^T$. For $c = (-1, -2, 0, 0)^T$, we write $c_B = (-1, -2)^T$ and $c_N = (0, 0)^T$. We let our starting vector be $\hat{x} = (20, 20, 0, 0)^T$, so that $\hat{x}_B = B^{-1}b = (20, 20)^T$, and $\hat{x}_N = (0, 0)^T$. Then we find that $y^T = c_B^T B^{-1} = (3, -1)$, and $y^T N = (-3, 1)$. The reduced cost vector is then

$$r^T = c_N^T - y^T N = (0, 0) - (-3, 1) = (3, -1).$$

Since r^T has a negative entry in its second position, $j = 2$, we learn that the entering variable is going to be $x_{2+j} = x_4$. The fourth column of A is $(0, -1)^T$, so the vector d^2 is

$$d^2 = B^{-1}(0, -1)^T = (1, -1)^T.$$

Therefore, we must select a new positive value for x_4 that satisfies

$$(20, 20) \geq x_4(1, -1).$$

The single positive entry of d^2 is the first one, from which we conclude that the leaving variable will be x_1. We therefore select as the new values of the variables $\hat{x}_1 = 0$, $\hat{x}_2 = 40$, $\hat{x}_3 = 0$, and $\hat{x}_4 = 20$. We then reorder the variables as $x = (x_4, x_2, x_3, x_1)^T$ and rearrange the columns of A accordingly. Having done this, we see that we now have

$$B = B_{\text{new}} = \begin{bmatrix} 0 & -1 \\ -1 & -1 \end{bmatrix},$$

with inverse

$$B^{-1} = \begin{bmatrix} 1 & -1 \\ -1 & 0 \end{bmatrix},$$

and the matrix N is

$$N = \begin{bmatrix} -1 & -1 \\ 0 & -2 \end{bmatrix}.$$

Since

$$B_{\text{new}} = B_{\text{old}} - \begin{bmatrix} -1 \\ -1 \end{bmatrix} \begin{bmatrix} 1 & 0 \end{bmatrix},$$

we can apply the Sherman–Morrison–Woodbury Identity to get B_{new}^{-1}.

The reduced cost vector is now $r^T = (2, 1)$. Since it has no negative entries, we have reached the optimal point; the solution is $\hat{x}_1 = 0$, $\hat{x}_2 = 40$, with slack variables $\hat{x}_3 = 0$ and $\hat{x}_4 = 20$.

6.13 Another Example

The following example is taken from Fang and Puthenpura [109]. Minimize the function

$$f(x_1, x_2, x_3, x_4, x_5, x_6) = -x_1 - x_2 - x_3,$$

subject to

$$2x_1 + x_4 = 1;$$

$$2x_2 + x_5 = 1;$$

$$2x_3 + x_6 = 1;$$

and $x_i \geq 0$, for $i = 1, ..., 6$. The variables x_4, x_5, and x_6 appear to be slack variables, introduced to obtain equality constraints.

Initially, we define the matrix A to be

$$A = \begin{bmatrix} 2 & 0 & 0 & 1 & 0 & 0 \\ 0 & 2 & 0 & 0 & 1 & 0 \\ 0 & 0 & 2 & 0 & 0 & 1 \end{bmatrix},$$

$b = (1, 1, 1)^T$, $c = (-1, -1, -1, 0, 0, 0)^T$ and $x = (x_1, x_2, x_3, x_4, x_5, x_6)^T$.

Suppose we begin with x_4, x_5, and x_6 as the basic variables. Since the entries of the vector b are positive, it is a simple matter to find an initial basic feasible solution; it is $x_4 = 1, x_5 = 1$, and $x_6 = 1$. We then rearrange the entries of the vector of unknowns so that

$$x = (x_4, x_5, x_6, x_1, x_2, x_3)^T.$$

Now we have to rearrange the columns of A as well; the new A is

$$A = \begin{bmatrix} 1 & 0 & 0 & 2 & 0 & 0 \\ 0 & 1 & 0 & 0 & 2 & 0 \\ 0 & 0 & 1 & 0 & 0 & 2 \end{bmatrix}.$$

The vector c must also be redefined; the new one is $c = (0, 0, 0, -1, -1, -1)^T$, so that $c_N = (-1, -1, -1)^T$ and $c_B = (0, 0, 0)^T$.

For this first step of the simplex method we have

$$B = \begin{bmatrix} 1 & 0 & 0 \\ 0 & 1 & 0 \\ 0 & 0 & 1 \end{bmatrix},$$

and

$$N = \begin{bmatrix} 2 & 0 & 0 \\ 0 & 2 & 0 \\ 0 & 0 & 2 \end{bmatrix}.$$

Note that one advantage in choosing the slack variables as the basic variables is that it is easy then to find the corresponding basic feasible solution, which is now

$$\hat{x} = \begin{bmatrix} \hat{x}_4 \\ \hat{x}_5 \\ \hat{x}_6 \\ \hat{x}_1 \\ \hat{x}_2 \\ \hat{x}_3 \end{bmatrix} = \begin{bmatrix} \hat{x}_B \\ \hat{x}_N \end{bmatrix} = \begin{bmatrix} 1 \\ 1 \\ 1 \\ 0 \\ 0 \\ 0 \end{bmatrix}.$$

The reduced cost vector r is then

$$r = (-1, -1, -1)^T;$$

since it has negative entries, the current basic feasible solution is not optimal.

Suppose that we select a non-basic variable with negative reduced cost, say x_1, which, we must remember, is the fourth entry of the redefined x, so $j = 1$ and $I + j = 4$. Then x_1 is the entering basic variable, and the vector d^1 is then

$$d^1 = B^{-1}a^j = (2, 0, 0)^T.$$

The only positive entry of d^1 is the first one, which means, according to Equation (6.4), that the exiting variable should be x_4. Now the new set of basic variables is $\{x_5, x_6, x_1\}$ and the new set of non-basic variables is $\{x_2, x_3, x_4\}$. The new matrices B and N are

$$B = \begin{bmatrix} 0 & 0 & 2 \\ 1 & 0 & 0 \\ 0 & 1 & 0 \end{bmatrix},$$

and

$$N = \begin{bmatrix} 0 & 0 & 1 \\ 2 & 0 & 0 \\ 0 & 2 & 0 \end{bmatrix}.$$

Continuing through two more steps, we find that the optimal solution is $-3/2$, and it occurs at the vector

$$x = (x_1, x_2, x_3, x_4, x_5, x_6)^T = (1/2, 1/2, 1/2, 0, 0, 0)^T.$$

6.14 Some Possible Difficulties

In the first example of the simplex method, we knew all four of the vertices of the feasible region, so we could choose any one of them to get our initial basic feasible solution. We chose to begin with x_1 and x_2 as our basic variables, which meant that the slack variables were zero and our first basic feasible solution was $\hat{x} = (20, 20, 0, 0)^T$. In the second example, we chose the slack variables to be the initial basic variables, which made it easy to find the initial basic feasible solution. Generally, however, finding an initial basic feasible solution may not be easy.

You might think that we can always simply take the slack variables as our initial basic variables, so that the initial B is just the identity matrix, and the initial basic feasible solution is merely the concatenation of the column vectors b and 0, as in the second example. The following example shows why this may not always work.

6.14.1 A Third Example

Consider the problem of minimizing the function $z = 2x_1 + 3x_2$, subject to

$$\begin{aligned} 3x_1 + 2x_2 &= 14, \\ 2x_1 - 4x_2 - x_3 &= 2, \\ 4x_1 + 3x_2 + x_4 &= 19, \text{ and} \\ x_i &\geq 0, \end{aligned}$$

for $i = 1, ..., 4$. The matrix A is now

$$A = \begin{bmatrix} 3 & 2 & 0 & 0 \\ 2 & -4 & -1 & 0 \\ 4 & 3 & 0 & 1 \end{bmatrix}.$$

There are only two slack variables, so we cannot construct our set of basic variables using only slack variables, since the matrix B must be square. We cannot begin with $\hat{x}_1 = \hat{x}_2 = 0$, since this would force $\hat{x}_3 = -2$, which is not permitted. We can choose $\hat{x}_2 = 0$ and solve for the other three, to get $\hat{x}_1 = \frac{14}{3}$, $\hat{x}_3 = \frac{22}{3}$, and $\hat{x}_4 = \frac{1}{3}$. This is relatively easy only because the problem is artificially small. The point here is that, for realistically large LP problems, finding a place to begin the simplex algorithm may not be a simple matter. For more on this matter, see [164].

In both of our first two examples, finding the inverse of the matrix B is easy, since B is only 2 by 2, or 3 by 3. In larger problems, finding B^{-1}, or better, solving $y^T B = c_B^T$ for y^T, is not trivial and can be an expensive part of each iteration. The Sherman–Morrison–Woodbury identity is helpful here.

6.15 Topics for Projects

The simplex method provides several interesting topics for projects.

(1) Investigate the issue of finding a suitable starting basic feasible solution. Reference [164] can be helpful in this regard.

(2) How can we reduce the cost associated with solving $y^T B = c_B^T$ for y^T at each step of the simplex method?

(3) Suppose that, instead of needing the variables to be nonnegative, we need each x_i to lie in the interval $[\alpha_i, \beta_i]$. How can we modify the simplex method to incorporate these constraints?

(4) Investigate the role of linear programming and the simplex method in graph theory and networks, with particular attention to the transport problem.

(5) There is a sizable literature on the computational complexity of the simplex method. Investigate this issue and summarize your findings.

6.16 Exercises

Ex. 6.1 *Prove Theorem 6.1 and its corollaries.*

Ex. 6.2 *Use Farkas' Lemma directly to prove that, if p^* is finite, then PS has a feasible solution.*

Ex. 6.3 *Put the Transport Problem into the form of an LP problem in DS form.*

Ex. 6.4 The Sherman–Morrison–Woodbury Identity *Let B be an invertible matrix. Show that*

$$(B - uv^T)^{-1} = B^{-1} + \alpha^{-1}(B^{-1}u)(v^T B^{-1}),$$

whenever

$$\alpha = 1 - v^T B^{-1} u \neq 0.$$

Show that, if $\alpha = 0$, then the matrix $B - uv^T$ has no inverse.

Ex. 6.5 *Show that B_{new} given in Equation (6.5) is invertible.*

Ex. 6.6 *Complete the calculation of the optimal solution for the problem in the second example of the simplex method.*

Ex. 6.7 *Consider the following problem, taken from [109]. Minimize the function*

$$f(x_1, x_2, x_3, x_4) = -3x_1 - 2x_2,$$

subject to

$$x_1 + x_2 + x_3 = 40,$$

$$2x_1 + x_2 + x_4 = 60,$$

and

$$x_j \geq 0,$$

for $j = 1, ..., 4$. *Use the simplex method to find the optimum solution. Take as a starting vector* $x^0 = (0, 0, 40, 60)^T$.

Ex. 6.8 *In the first example on the simplex method, the new value of* x_2 *became* 40. *Explain why this was the case.*

Ex. 6.9 *Redo the first example of the simplex method, starting with the vertex* $x_1 = 0$ *and* $x_2 = 0$.

Ex. 6.10 *Consider the LP problem of maximizing the function* $f(x_1, x_2) = x_1 + 2x_2$, *subject to*

$$\begin{aligned} -2x_1 + x_2 &\leq 2, \\ -x_1 + 2x_2 &\leq 7, \\ x_1 &\leq 3, \\ x_1 &\geq 0, \text{ and} \\ x_2 &\geq 0. \end{aligned}$$

Start at $x_1 = 0$, $x_2 = 0$. *You will find that you have a choice for the entering variable; try it both ways.*

Ex. 6.11 *Carry out the next two steps of the simplex algorithm for the second example given earlier.*

Ex. 6.12 *Apply the simplex method to minimize* $z = -x_1 - 2x_2$, *subject to*

$$\begin{aligned} -x_1 + x_2 &\leq 2, \\ -2x_1 + x_2 &\leq 1, \\ x_1 &\geq 0, \text{ and} \\ x_2 &\geq 0. \end{aligned}$$

Chapter 7

Matrix Games and Optimization

7.1 Chapter Summary

All the optimization problems discussed so far have involved a single individual trying to maximize or minimize some function. In 1928 the mathematician John von Neumann introduced the theory of games in an attempt to deal with the more general problem of two or more individuals in competition and to make economics more scientific. The theory of games was developed somewhat later by von Neumann and Morgenstern [167] and has become an important tool, not only in economics, but throughout the social sciences. Two-person zero-sum games provide a nice example of optimization and an opportunity to apply some of the linear algebra and linear programming tools previously discussed. In this chapter we introduce the idea of two-person matrix games and use results from linear programming to prove von Neumann's "Minimax Theorem," also called the Fundamental Theorem of Game Theory. Our focus here is on the mathematics; the DVD course by Stevens [193] provides a less mathematical introduction to game theory, with numerous examples drawn from business and economics. The classic book by Schelling [184] describes the roles played by game theory in international politics and warfare.

7.2 Two-Person Zero-Sum Games

A two-person game is called a *constant-sum game* if the total payout is the same, each time the game is played. In such cases, we can subtract half the total payout from the payout to each player and record only the difference. Then the total payout appears to be zero, and such games are called *zero-sum games*. We can then suppose that whatever one player wins is paid by the other player. Except for the final section, we shall consider only two-person, zero-sum games.

7.3 Deterministic Solutions

In this two-person game, the first player, call him P1, selects a row of the I by J real matrix A, say i, and the second player selects a column of A, say j. The second player, call her P2, pays the first player A_{ij}. If some $A_{ij} < 0$, then this means that the first player pays the second. Since whatever the first player wins, the second loses, and vice versa, we need only one matrix to summarize the situation. Note that, even though we label the players in order, their selections are made simultaneously and without knowledge of the other player's selection.

7.3.1 Optimal Pure Strategies

In our first example, the matrix is

$$A = \begin{bmatrix} 7 & 8 & 4 \\ 4 & 7 & 2 \end{bmatrix}. \tag{7.1}$$

The first player notes that by selecting row $i = 1$, he will get at least 4, regardless of which column the second player plays. The second player notes that, by playing column $j = 3$, she will pay the first player no more than 4, regardless of which row the first player plays. If the first player then begins to play $i = 1$ repeatedly, and the second player notices this consistency, she will still have no motivation to play any column except $j = 3$, because the other payouts are both worse than 4. Similarly, so long as the second player is playing $j = 3$ repeatedly, the first player has no motivation to play anything other than $i = 1$, since he will be paid less if he switches. Therefore, both players adopt a *pure strategy* of $i = 1$ and $j = 3$. This game is said to be *deterministic* and the entry $A_{1,3} = 4$ is a *saddle point* because it is the maximum of its column and the minimum of its row.

Note that we can write

$$A_{i,3} \leq A_{1,3} \leq A_{1,j},$$

so we have

$$\max_i \min_j A_{ij} = 4 = \min_j \max_i A_{ij}.$$

Once the two players play $(1, 3)$ neither has any motivation to change. For this reason the entry $A_{1,3}$ is called a Nash equilibrium. The value $A_{1,3} = 4$ is the maximum of the minimum wins the first player can have, and also the minimum of the maximum losses the second player can suffer. Not all such two-person games have saddle points, however.

Consider now the two-person game with payoff matrix

$$B = \begin{bmatrix} 4 & 1 \\ 2 & 3 \end{bmatrix}. \tag{7.2}$$

Unlike the matrix A in Equation (7.1), the matrix B in Equation (7.2) has no saddle point; no entry of B is the maximum of its column and the minimum of its row. For such games we need to use randomized strategies.

7.4 Randomized Solutions

When the game has no saddle point, there is no optimal deterministic solution. Instead, we consider approaches that involve selecting our strategies according to some random procedure, and seek an optimal randomized strategy.

7.4.1 Optimal Randomized Strategies

Consider the game described by the matrix B in Equation (7.2). The first player notes that by selecting row $i = 2$, he will get at least 2, regardless of which column the second player plays. The second player notes that, by playing column $j = 2$, she will pay the first player no more than 3, regardless of which row the first player plays. If both begin by playing in this conservative manner, the first player will play $i = 2$ and the second player will play $j = 2$.

If the first player plays $i = 2$ repeatedly, and the second player notices this consistency, she will be tempted to switch to playing column $j = 1$, thereby losing only 2, instead of 3. If she makes the switch and the first player notices, he will be motivated to switch his play to row $i = 1$, to get a payoff of 4, instead of 2. The second player will then soon switch to playing $j = 2$ again, hoping that the first player sticks with $i = 1$. But the first player is not stupid, and quickly returns to playing $i = 2$. There is no saddle point in this game; the maximum of the minimum wins the first player can have is 2, but the minimum of the maximum losses the second player can suffer is 3. For such games, it makes sense for both players to select their play at random, with the first player playing $i = 1$ with probability p and $i = 2$ with probability $1 - p$, and the second player playing column $j = 1$ with probability q and $j = 2$ with probability $1 - q$. These are called *randomized strategies*.

When the first player plays $i = 1$, he expects to get $4q+(1-q) = 3q+1$, and when he plays $i = 2$ he expects to get $2q + 3(1 - q) = 3 - q$. Note that $3q + 1 = 3 - q$ when $q = 0.5$, so if the second player plays $q = 0.5$, then the second player will not care what the first player does, since the expected payoff to the first player is $5/2$ in either case. If the second player plays a different q, then the payoff to the first player will depend on what the first player does, and can be larger than $5/2$.

Since the first player plays $i = 1$ with probability p, he expects to get

$$p(3q + 1) + (1 - p)(3 - q) = 4pq - 2p - q + 3 = (4p - 1)q + 3 - 2p.$$

He notices that if he selects $p = \frac{1}{4}$, then he expects to get $\frac{5}{2}$, regardless of what the second player does. If he plays something other than $p = \frac{1}{4}$, his expected winnings will depend on what the second player does. If he selects a value of p less than $\frac{1}{4}$, and $q = 1$ is selected, then he wins $2p + 2$, but this is less than $\frac{5}{2}$. If he selects $p > \frac{1}{4}$ and $q = 0$ is selected, then he wins $3 - 2p$, which again is less than $\frac{5}{2}$. The maximum of these minimum payoffs occurs when $p = \frac{1}{4}$ and the *max-min* win is $\frac{5}{2}$.

Similarly, the second player, noticing that

$$p(3q + 1) + (1 - p)(3 - q) = (4q - 2)p + 3 - q,$$

sees that she will pay out $\frac{5}{2}$ if she takes $q = \frac{1}{2}$. If she selects a value of q

less than $\frac{1}{2}$, and $p = 0$ is selected, then she pays out $3 - q$, which is more than $\frac{5}{2}$. If, on the other hand, she selects a value of q that is greater than $\frac{1}{2}$, and $p = 1$ is selected, then she will pay out $3q+1$, which again is greater than $\frac{5}{2}$. The only way she can be certain to pay out no more than $\frac{5}{2}$ is to select $q = \frac{1}{2}$. The minimum of these maximum payouts occurs when she chooses $q = \frac{1}{2}$, and the *min-max* payout is $\frac{5}{2}$. The choices of $p = \frac{1}{4}$ and $q = \frac{1}{2}$ constitute a Nash equilibrium, because, once these choices are made, neither player has any reason to change strategies.

This leads us to the question of whether or not there will always be probability vectors for the players that will lead to the equality of the max-min win and the min-max payout.

We make a notational change at this point. From now on the letters p and q will denote probability column vectors, and not individual probabilities, as previously. Note that, in general, since $A_{i,j}$ is the payout to P1 when (i, j) is played, for $i = 1, ..., I$ and $j = 1, ..., J$, and the probability that (i, j) will be played is $p_i q_j$, the expected payout to P1 is

$$\sum_{i=1}^{I} \sum_{j=1}^{J} p_i A_{i,j} q_j = p^T A q.$$

The probabilities $\hat{p}$ and $\hat{q}$ will be optimal randomized strategies if

$$p^T A \hat{q} \leq \hat{p}^T A \hat{q} \leq \hat{p}^T A q,$$

for any probabilities p and q. Once again, we have a Nash equilibrium, since once the optimal strategies are the chosen ones, neither player has any motivation to adopt a different randomized strategy.

7.4.2 An Exercise

Ex. 7.1 *Suppose that there are two strains of flu virus and two types of vaccine. The first vaccine, call it V1, is* 0.85 *effective against the first strain (F1) and* 0.70 *effective against the second (F2), while the second vaccine (V2) is* 0.60 *effective against F1 and* 0.90 *effective against F2. The public health service is the first player, P1, and nature is the second player, P2. The service has to decide what percentage of the vaccines manufactured and made available to the public are to be of type V1 and what percentage are to be of type V2, while not knowing what percentage of the flu virus is F1 and what percentage is F2. Set this up as a matrix game and determine how the public health service should proceed.*

7.4.3 The Min-Max Theorem

Let A be an I by J payoff matrix. Let

$$P = \Big\{p = (p_1, ..., p_I)^T \,|\, p_i \geq 0, \sum_{i=1}^{I} p_i = 1\Big\},$$

$$Q = \Big\{q = (q_1, ..., q_J)^T \,|\, q_j \geq 0, \sum_{j=1}^{J} q_j = 1\Big\},$$

and

$$R = A(Q) = \{Aq \,|q \in Q\}.$$

The first player selects a vector p in P and the second selects a vector q in Q. The expected payoff to the first player is

$$E = \langle p, Aq \rangle \ = p^T Aq.$$

Let

$$m_0 = \max_{p \in P} \min_{r \in R} \langle p, r \rangle,$$

and

$$m^0 = \min_{r \in R} \max_{p \in P} \langle p, r \rangle;$$

the interested reader may want to prove that the maximum and minimum exist. Clearly, we have

$$\min_{r \in R} \langle p, r \rangle \leq \langle p, r \rangle \leq \max_{p \in P} \langle p, r \rangle,$$

for all $p \in P$ and $r \in R$. It follows that $m_0 \leq m^0$. The Min-Max Theorem, also known as the Fundamental Theorem of Game Theory, asserts that $m_0 = m^0$.

Theorem 7.1 The Fundamental Theorem of Game Theory *Let A be an arbitrary real I by J matrix. Then there are vectors $\hat{p}$ in P and $\hat{q}$ in Q such that*

$$p^T A\hat{q} \leq \hat{p}^T A\hat{q} \leq \hat{p}^T Aq, \tag{7.3}$$

for all p in P and q in Q.

The quantity $\omega = \hat{p}^T A\hat{q}$ is called the *value of the game*. Notice that if P1 knows that P2 plays according to the mixed-strategy vector q, P1 could examine the entries $(Aq)_i$, which are his expected payoffs should he play strategy i, and select the one for which this expected payoff is largest. However, if P2 notices what P1 is doing, she can abandon q to her advantage. When $q = \hat{q}$, it follows, from the inequalities in (7.3) by using p with the ith entry equal to one and the rest zero, that

$$(A\hat{q})_i \leq \omega$$

for all i, and

$$(A\hat{q})_i = \omega$$

for all i for which $\hat{p}_i > 0$. So there is no long-term advantage to P1 to move away from $\hat{p}$.

There are a number of different proofs of the Fundamental Theorem, including one using Fenchel Duality. In this chapter we consider proofs based on linear-algebraic methods, linear programming, and theorems of the alternative.

7.5 Symmetric Games

A game is said to be *symmetric* if the available strategies are the same for both players, and if the players switch strategies, the outcomes switch also. In other words, the payoff matrix A is skew-symmetric, that is, A is square and $A_{ji} = -A_{ij}$. For symmetric games, we can use Theorem 4.12 to prove the existence of a randomized solution.

First, we show that there is a probability vector $\hat{p} \geq 0$ such that $\hat{p}^T A \geq 0$. Then we show that

$$p^T A\hat{p} \leq 0 = \hat{p}^T A\hat{p} \leq \hat{p}^T Aq,$$

for all probability vectors p and q. It will then follow that $\hat{p}$ and $\hat{q} = \hat{p}$ are the optimal mixed strategies.

If there is no nonzero $x \geq 0$ such that $x^T A \geq 0$, then there is no nonzero $x \geq 0$ such that $A^T x \geq 0$. Then, by Theorem 4.12, we know that there is $y \geq 0$ with $Ay < 0$; obviously y is not the zero vector, in this case. Since $A^T = -A$, it follows that $y^T A > 0$. Consequently, there is a nonzero $x \geq 0$, such that $x^T A \geq 0$; it is $x = y$. This is a contradiction. So $\hat{p}$ exists.

Since the game is symmetric, we have

$$\hat{p}^T A\hat{p} = (\hat{p}^T A\hat{p})^T = \hat{p}^T A^T \hat{p} = -\hat{p}^T A\hat{p},$$

so that $\hat{p}^T A\hat{p} = 0$.

For any probability vectors p and q we have

$$p^T A\hat{p} = \hat{p}^T A^T p = -\hat{p}^T Ap \leq 0,$$

and

$$0 \leq \hat{p}^T Aq.$$

We conclude that the mixed strategies $\hat{p}$ and $\hat{q} = \hat{p}$ are optimal.

7.5.1 An Example of a Symmetric Game

We present now a simple example of a symmetric game and compute the optimal randomized strategies.

Consider the payoff matrix

$$A = \begin{bmatrix} 0 & 1 \\ -1 & 0 \end{bmatrix}.$$

This matrix is skew-symmetric, so the game is symmetric. Let $\hat{p}^T = [1, 0]$; then $\hat{p}^T A = [0, 1] \geq 0$. We show that $\hat{p}$ and $\hat{q} = \hat{p}$ are the optimal randomized strategies. For any probability vectors $p^T = [p_1, p_2]$ and $q^T = [q_1, q_2]$, we have

$$p^T A\hat{p} = -p_2 \leq 0,$$

$$\hat{p}^T A\hat{p} = 0,$$

and

$$\hat{p}^T Aq = q_2 \geq 0.$$

It follows that the pair of strategies $\hat{p} = \hat{q} = [1, 0]^T$ are optimal randomized strategies.

7.5.2 Comments on the Proof of the Min-Max Theorem

In [115], Gale proves the existence of optimal randomized solutions for an arbitrary matrix game by showing that there is associated with such a game a symmetric matrix game and that an optimal randomized solution exists for one if and only if such exists for the other. Another way is by converting the existing game into a "positive" game.

7.6 Positive Games

As Gale notes in [115], it is striking that two fundamental mathematical tools in linear economic theory, linear programming and game theory, developed simultaneously, and independently, in the years following the Second World War. More remarkable still was the realization that these two areas are closely related. Gale's proof of the Min-Max Theorem, which relates the game to a linear programming problem and employs his Strong Duality Theorem, provides a good illustration of this close connection.

If the I by J payoff matrix A has only positive entries, we can use Gale's Strong Duality Theorem 6.3 for linear programming to prove the Min-Max Theorem.

Let b and c be the vectors whose entries are all one. Consider the LP problem of minimizing $z = c^T x$, over all $x \geq 0$ with $A^T x \geq b$; this is the PC problem. The DC problem is then to maximize $w = b^T y$, over all $y \geq 0$ with $Ay \leq c$. Since A has only positive entries, both PC and DC are feasible, so, by Gale's Strong Duality Theorem 6.3, we know that there are feasible nonnegative vectors $\hat{x}$ and $\hat{y}$ and nonnegative μ such that

$$\hat{z} = c^T \hat{x} = \mu = b^T \hat{y} = \hat{w}.$$

Since $\hat{x}$ cannot be zero, μ must be positive.

7.6.1 Some Exercises

Ex. 7.2 *Show that the vectors $\hat{p} = \frac{1}{\mu}\hat{x}$ and $\hat{q} = \frac{1}{\mu}\hat{y}$ are probability vectors and are optimal randomized strategies for the matrix game.*

Ex. 7.3 *Given an arbitrary I by J matrix A, there is $\alpha > 0$ so that the matrix B with entries $B_{ij} = A_{ij} + \alpha$ has only positive entries. Show that any optimal randomized probability vectors for the game with payoff matrix B are also optimal for the game with payoff matrix A.*

It follows from these exercises that there exist optimal randomized solutions for any matrix game.

7.6.2 Comments

This proof of the Min-Max Theorem shows that we can associate with a given matrix game a linear programming problem. It follows that we can use the simplex method to find optimal randomized solutions for matrix games. It also suggests that a given linear programming problem can be associated with a matrix game; see Gale [115] for more discussion of this point.

7.7 Example: The "Bluffing" Game

In [115] Gale discusses several games, one of which he calls the "bluffing" game. For this game, there is a box containing two cards, marked HI and LO, respectively. Both players begin by placing their "ante" $a > 0$, on the table. Player One, P1, draws one of the two cards and looks at it; Player Two, P2, does not see it. Then P1 can either "fold," losing his ante $a > 0$ to P2, or "bet" $b > a$. Then P2 can either fold, losing her ante also to P1, or "call," and bet b also. If P2 calls, she wins if LO is on the card drawn, and P1 wins if it is HI.

Since it makes no sense for P1 to fold when HI, his two strategies are

- s1: bet in both cases; and
- s2: bet if HI and fold if LO.

Strategy s1 is "bluffing" on the part of P1, since he bets even when he knows the card shows LO.

Player Two has the two strategies

- t1: call; and
- t2: fold.

When (s1,t1) is played, P1 wins the bet half the time, so his expected gain is zero.

When (s1,t2) is played, P1 wins the ante a from P2.

When (s2,t1) is played, P1 bets half the time, winning each time, so gaining b, but loses his ante a half the time. His expected gain is then $(b-a)/2$.

When (s2,t2) is played, P1 wins the ante from P2 half the time, and they exchange antes half the time. Therefore, P1 expects to win $a/2$.

The payoff matrix for P1 is then

$$A = \begin{bmatrix} 0 & a \\ \frac{b-a}{2} & \frac{a}{2} \end{bmatrix}.$$

Note that if $b \leq 2a$, then the game has a saddle point, (s2,t1), and the saddle value is $\frac{b-a}{2}$. If $b > 2a$, then the players need randomized strategies.

Suppose P1 plays s1 with probability p and s2 with probability $1-p$, while P2 plays t1 with probability q and t2 with probability $1-q$. Then the expected gain for P1 is

$$p(1-q)a + (1-p)\Big(q\frac{b-a}{2} + (1-q)\frac{a}{2}\Big),$$

which can be written as

$$(1+p)\frac{a}{2} + q\Big((1-p)\frac{b}{2} - a\Big),$$

and as

$$\frac{a}{2} + q\Big(\frac{b}{2} - a\Big) + p\Big(\frac{a}{2} - q\frac{b}{2}\Big).$$

If

$$\Big((1-p)\frac{b}{2} - a\Big) = 0,$$

or $p = 1 - \frac{2a}{b}$, then P1 expects to win

$$a - \frac{a^2}{b} = \frac{2a}{b}\frac{b-a}{2},$$

regardless of what q is. Similarly, if

$$\left(\frac{a}{2} - q\frac{b}{2}\right) = 0,$$

or $q = \frac{a}{b}$, then P2 expects to pay out $a - \frac{a^2}{b}$, regardless of what p is. These are the optimal randomized strategies.

If $b \leq 2a$, then P1 should never bluff, and should always play s2. Then P2 will always play t1 and P1 wins $\frac{b-a}{2}$, on average. But when b is higher than $2a$, P2 would always play t2, if P1 always plays s2, in which case the payoff would be only $\frac{a}{2}$, which is lower than the expected payoff when P1 plays optimally. It pays P1 to bluff, because it forces P2 to play t1 some of the time.

7.8 Learning the Game

In our earlier discussion we saw that the matrix game involving the payoff matrix

$$A = \begin{bmatrix} 4 & 1 \\ 2 & 3 \end{bmatrix}$$

is not deterministic. The best thing the players can do is to select their play at random, with the first player playing $i = 1$ with probability p and $i = 2$ with probability $1 - p$, and the second player playing column $j = 1$ with probability q and $j = 2$ with probability $1 - q$. If the first player, call him P1, selects $p = \frac{1}{4}$, then he expects to get $\frac{5}{2}$, regardless of what the second player, call her P2, does; otherwise his fortunes depend on what P2 does. His optimal mixed-strategy (column) vector is $[1/4, 3/4]^T$. Similarly, the second player notices that the only way she can be certain to pay out no more than $\frac{5}{2}$ is to select $q = \frac{1}{2}$. The minimum of these maximum payouts occurs when she chooses $q = \frac{1}{2}$, and the *min-max* payout is $\frac{5}{2}$.

Because the payoff matrix is two-by-two, we are able to determine easily the optimal mixed-strategy vectors for each player. When the payoff matrix is larger, finding the optimal mixed-strategy vectors is not a simple matter. As we have seen, one approach is to obtain these vectors by solving a related linear-programming problem. In this section we consider other approaches to finding the optimal mixed-strategy vectors.

7.8.1 An Iterative Approach

In [115] Gale presents an iterative approach to learning how best to play a matrix game. The assumptions are that the game is to be played repeatedly and that the two players adjust their play as they go along, based on the earlier plays of their opponent.

Suppose, for the moment, that P1 knows that P2 is playing the randomized strategy q, where, as earlier, we denote by p and q probability column vectors. The entry $(Aq)_i$ of the column vector Aq is the expected payoff to P1 if he plays strategy i. It makes sense for P1 then to find the index i for which this expected payoff is largest and to play that strategy every time. Of course, if P2 notices what P1 is doing, she will abandon q to her advantage.

After the game has been played n times, the players can examine the previous plays and make estimates of what the opponent is doing. Suppose that P1 has played strategy i n_i times, where $n_i \geq 0$ and $n_1 + n_2 + ... + n_I = n$. Denote by p^n the probability column vector whose ith entry is n_i/n. Similarly, calculate q^n. These two probability vectors summarize the tendencies of the two players over the first n plays. It seems reasonable that an attempt to learn the game would involve these probability vectors.

For example, P1 could see which entry of q^n is the largest, assume that P2 is most likely to play that strategy the next time, and play his best strategy against that play of P2. However, if there are several strategies for P2 to choose, it is still unlikely that P2 will choose this strategy the next time. Perhaps P1 could do better by considering his long-run fortunes and examining the vector Aq^n of expected payoffs. In the exercise below, you are asked to investigate this matter.

7.8.2 An Exercise

As we have seen, the optimal randomized strategies can be found by solving a linear programming problem. The following exercise suggests a different way to discover these strategies.

Ex. 7.4 *Suppose that both players are attempting to learn how best to play the game by examining the vectors p^n and q^n after n plays. Devise an algorithm for the players to follow that will lead to optimal mixed strategies for both. Simulate repeated play of a particular matrix game to see how your algorithm performs. If the algorithm does its job, but does it slowly, that is, it takes many plays of the game for it to begin to work, investigate how it might be speeded up.*

7.9 Non-Constant-Sum Games

In this final section we consider non-constant-sum games. These are more complicated and the mathematical results more difficult to obtain than in the constant-sum games. Such non-constant-sum games can be used to model situations in which the players may both gain by cooperation, or, when speaking of economic actors, by collusion [99]. We begin with the most famous example of a non-constant-sum game, the Prisoners' Dilemma.

7.9.1 The Prisoners' Dilemma

Imagine that you and your partner are arrested for robbing a bank and both of you are guilty. The two of you are held in separate rooms and given the following options by the district attorney: (1) if you confess, but your partner does not, you go free, while he gets three years in jail; (2) if he confesses, but you do not, he goes free and you get the three years; (3) if both of you confess, you each get two years; (4) if neither of you confesses, each of you gets one year in jail. Let us call you player number one, and your partner player number two. Let strategy one be to remain silent, and strategy two be to confess.

Your payoff matrix is

$$A = \begin{bmatrix} -1 & -3 \\ 0 & -2 \end{bmatrix},$$

so that, for example, if you remain silent, while your partner confesses, your payoff is $A_{1,2} = -3$, where the negative sign is used because jail time is undesirable. From your perspective, the game has a deterministic solution; you should confess, assuring yourself of no more than two years in jail. Your partner views the situation the same way and also should confess. However, when the game is viewed, not from one individual's perspective, but from the perspective of the pair of you, we see that by sticking together you each get one year in jail, instead of each of you getting two years; if you cooperate, you both do better.

7.9.2 Two Payoff Matrices Needed

In the case of non-constant-sum games, one payoff matrix is not enough to capture the full picture. Consider the following example of a non-constant-sum game. Let the matrix

$$A = \begin{bmatrix} 5 & 4 \\ 3 & 6 \end{bmatrix}$$

be the payoff matrix for Player One (P_1), and

$$B = \begin{bmatrix} 5 & 6 \\ 7 & 2 \end{bmatrix}$$

be the payoff matrix for Player Two (P_2); that is, $A_{1,2} = 4$ and $B_{2,1} = 7$ means that if P_1 plays the first strategy and P_2 plays the second strategy, then P_1 gains four and P_2 gains seven. Notice that the total payoff for each play of the game is not constant, so we require two matrices, not one.

Player One, considering only the payoff matrix A, discovers that the best strategy is a randomized strategy, with the first strategy played three quarters of the time. Then P_1 has expected gain of $\frac{9}{2}$. Similarly, Player Two, applying the same analysis to his payoff matrix, B, discovers that he should also play a randomized strategy, playing the first strategy five sixths of the time; he then has an expected gain of $\frac{16}{3}$. However, if P_1 switches and plays the first strategy all the time, while P_2 continues with his randomized strategy, P_1 expects to gain $\frac{29}{6} > \frac{27}{6}$, while the expected gain of P_2 is unchanged. This is very different from what happens in the case of a constant-sum game; there, the sum of the expected gains is constant, and equals zero for a zero-sum game, so P_1 would not be able to increase his expected gain, if P_2 plays his optimal randomized strategy.

7.9.3 An Example: Illegal Drugs in Sports

In a recent article in Scientific American [188], Michael Shermer uses the model of a non-constant-sum game to analyze the problem of doping, or illegal drug use, in sports, and to suggest a solution. He is a former competitive cyclist and his specific example comes from the Tour de France. He is the first player, and his opponent the second player. The choices are to cheat by taking illegal drugs or to stay within the rules. The assumption he makes is that a cyclist who sticks to the rules will become less competitive and will be dropped from his team.

Currently, the likelihood of getting caught is low, and the penalty for cheating is not too high, so, as he shows, the rational choice is for everyone to cheat, as well as for every cheater to lie. He proposes changing the payoff matrices by increasing the likelihood of being caught, as well as the penalty for cheating, so as to make sticking to the rules the rational choice.

Chapter 8

Differentiation

8.1 Chapter Summary

The definition of the derivative of a function $g : D \subseteq \mathbb{R} \to \mathbb{R}$ is a familiar one. In this chapter we examine various ways in which this definition can be extended to functions $f : D \subseteq \mathbb{R}^J \to \mathbb{R}$ of several variables. Here D is the domain of the function f and we assume that $\text{int}(D)$, the interior of the set D, is not empty. While the concepts of directional derivatives and gradients are familiar enough, they are not the whole story of differentiation. In this chapter we consider the Gâteaux derivative and the Fréchet derivative, along with several examples. This chapter can be skipped without harm to the reader.

8.2 Directional Derivative

We begin with one- and two-sided directional derivatives.

8.2.1 Definitions

The function $g(x) = |x|$ does not have a derivative at $x = 0$, but it has *one-sided directional derivatives* there. The one-sided directional derivative of $g(x)$ at $x = 0$, in the direction of $x = 1$, is

$$g'_+(0;1) = \lim_{t \downarrow 0} \frac{1}{t}[g(0+t) - g(0)] = 1,$$

and in the direction of $x = -1$, it is

$$g'_+(0;-1) = \lim_{t \downarrow 0} \frac{1}{t}[g(0-t) - g(0)] = 1.$$

However, the two-sided derivative of $g(x) = |x|$ does not exist at $x = 0$.

We can extend the concept of one-sided directional derivatives to functions of several variables.

Definition 8.1 *Let $f : D \subseteq \mathbb{R}^J \to \mathbb{R}$ be a real-valued function of several variables, let a be in int(D), and let d be a unit vector in $\mathbb{R}^J$. The* one-sided directional derivative *of $f(x)$, at $x = a$, in the direction of d, is*

$$f'_+(a;d) = \lim_{t \downarrow 0} \frac{1}{t}[f(a+td) - f(a)].$$

Definition 8.2 *The* two-sided directional derivative *of $f(x)$ at $x = a$, in the direction of d, is*

$$f'(a;d) = \lim_{t \to 0} \frac{1}{t}(f(a+td) - f(a)).$$

If the two-sided directional derivative exists then we have

$$f'(a;d) = f'_+(a;d) = -f'_+(a;-d).$$

Given $x = a$ and d, we define the function $\phi(t) = f(a+td)$, for t such that $a + td$ is in D. The derivative of $\phi(t)$ at $t = 0$ is then

$$\phi'(0) = \lim_{t \to 0} \frac{1}{t}[\phi(t) - \phi(0)] = f'(a;d).$$

In the definition of $f'(a;d)$ we restricted d to unit vectors because the directional derivative $f'(a;d)$ is intended to measure the rate of change of $f(x)$ as x moves away from $x = a$ in the direction d. Later, in our discussion of convex functions, it will be convenient to view $f'(a;d)$ as a function of d and to extend this function to the more general function of arbitrary z defined by

$$f'(a;z) = \lim_{t \to 0} \frac{1}{t}(f(a+tz) - f(a)).$$

It is easy to see that

$$f'(a;z) = \|z\|_2 f'(a; z/\|z\|_2).$$

8.3 Partial Derivatives

For $j = 1, ..., J$, denote by e^j the vector whose entries are all zero, except for a one in the jth position.

Definition 8.3 *If $f'(a; e^j)$ exists, then it is $\frac{\partial f}{\partial x_j}(a)$, the* partial derivative *of $f(x)$, at $x = a$, with respect to x_j, the jth entry of the variable vector x.*

Definition 8.4 *If the partial derivative, at $x = a$, with respect to x_j, exists for each j, then the* gradient *of $f(x)$, at $x = a$, is the vector $\nabla f(a)$ whose entries are $\frac{\partial f}{\partial x_j}(a)$.*

8.4 Some Examples

We consider some examples of directional derivatives.

8.4.1 Example 1

For $(x, y) \neq (0, 0)$, let

$$f(x, y) = \frac{2xy}{x^2 + y^2},$$

and define $f(0, 0) = 1$. Let $d = (\cos\theta, \sin\theta)$. Then it is easy to show that, for $a = (0, 0)$,

$$\phi(t) = f(a + td) = \sin 2\theta,$$

for $t \neq 0$, and $\phi(0) = 1$. If θ is such that $\sin 2\theta = 1$, then $\phi(t)$ is constant, and $\phi'(0) = 0$. But, if $\sin 2\theta \neq 1$, then $\phi(t)$ is discontinuous at $t = 0$, so $\phi(t)$ is not differentiable at $t = 0$. Therefore, $f(x, y)$ has a two-sided directional derivative at $(x, y) = (0, 0)$ only in certain directions.

8.4.2 Example 2

[**114**] For $(x, y) \neq (0, 0)$, let

$$f(x, y) = \frac{2xy^2}{x^2 + y^4},$$

and $f(0,0) = 0$. Again, let $d = (\cos\theta, \sin\theta)$. Then we have

$$\phi'(0) = \frac{2\sin^2\theta}{\cos\theta},$$

for $\cos\theta \neq 0$. If $\cos\theta = 0$, then $f(x,y)$ is the constant zero in that direction, so $\phi'(0) = 0$. Therefore, the function $f(x,y)$ has a two-sided directional derivative at $(x,y) = (0,0)$, for every vector d. Note that the two partial derivatives are both zero at $(x,y) = (0,0)$, so $\nabla f(0,0) = 0$. Note also that, since $f(y^2, y) = 1$ for all $y \neq 0$, the function $f(x,y)$ is not continuous at $(0,0)$.

8.5 Gâteaux Derivative

Just having a two-sided directional derivative for every d is not sufficient, in most cases; we need something stronger.

Definition 8.5 *If $f(x)$ has a two-sided directional derivative at $x = a$, for every vector d, and, in addition,*

$$f'(a; d) = \langle \nabla f(a), d \rangle,$$

for each d, then $f(x)$ is Gâteaux-differentiable *at $x = a$, and $\nabla f(a)$ is the* Gâteaux derivative *of $f(x)$ at $x = a$, also denoted $f'(a)$.*

Example 2 above showed that it is possible for $f(x)$ to have a two-sided directional derivative at $x = a$, for every d, and yet fail to be Gâteaux-differentiable.

From Cauchy's Inequality, we know that

$$|f'(a; d)| = |\langle \nabla f(a), d \rangle| \leq ||\nabla f(a)||_2 \, ||d||_2,$$

and that $f'(a; d)$ attains its most positive value when the direction d is a positive multiple of $\nabla f(a)$. This is the motivation for steepest descent optimization.

For ordinary functions $g : D \subseteq \mathbb{R} \to \mathbb{R}$, we know that differentiability implies continuity. It is possible for $f(x)$ to be Gâteaux-differentiable at $x = a$ and yet not be continuous at $x = a$; see Ortega and Rheinboldt [174]. This means that the notion of Gâteaux-differentiability is too weak. In order to have a nice theory of multivariate differentiation, the notion of derivative must be strengthened. The stronger notion we seek is Fréchet differentiability.

8.6 Fréchet Derivative

The notion of Fréchet-differentiability is the one appropriate for our purposes.

8.6.1 The Definition

Definition 8.6 *We say that $f(x)$ is* Fréchet-differentiable *at $x = a$ and $\nabla f(a)$ is its Fréchet derivative if*

$$\lim_{||h|| \to 0} \frac{1}{||h||} |f(a+h) - f(a) - \langle \nabla f(a), h \rangle| = 0.$$

Notice that the limit in the definition of the Fréchet derivative involves the norm of the incremental vector h, which is where the power of the Fréchet derivative arises. Also, since the norm and the associated inner product can be changed, so can the Fréchet derivative; see Exercise 8.1 for an example. The corresponding limit in the definition of the Gâteaux derivative involves only the scalar t, and therefore requires no norm and makes sense in any vector space.

8.6.2 Properties of the Fréchet Derivative

It can be shown that if $f(x)$ is Fréchet-differentiable at $x = a$, then $f(x)$ is continuous at $x = a$. If $f(x)$ is Gâteaux-differentiable at each point in an open set containing $x = a$, and $\nabla f(x)$ is continuous at $x = a$, then $\nabla f(a)$ is also the Fréchet derivative of $f(x)$ at $x = a$. Since the continuity of $\nabla f(x)$ is equivalent to the continuity of each of the partial derivatives, we learn that $f(x)$ is Fréchet-differentiable at $x = a$ if it is Gâteaux-differentiable in a neighborhood of $x = a$ and the partial derivatives are continuous at $x = a$. If $\nabla f(x)$ is continuous in a neighborhood of $x = a$, the function $f(x)$ is said to be *continuously differentiable*. Unless we write otherwise, when we say that a function is differentiable, we shall mean Gâteaux-differentiable, since this is usually sufficient for our purposes and the two types of differentiability typically coincide anyway.

8.7 The Chain Rule

For fixed a and d in $\mathbb{R}^J$, the function $\phi(t) = f(a + td)$, defined for the real variable t, is a composition of the function $f : \mathbb{R}^J \to \mathbb{R}$ itself and the function $g : \mathbb{R} \to \mathbb{R}^J$ defined by $g(t) = a + td$; that is, $\phi(t) = f(g(t))$. Writing

$$f(a + td) = f(a_1 + td_1, a_2 + td_2, ..., a_J + td_J),$$

and applying the Chain Rule, we find that

$$f'(a; d) = \phi'(0) = \frac{\partial f}{\partial x_1}(a)d_1 + ... + \frac{\partial f}{\partial x_J}(a)d_J;$$

that is,

$$f'(a; d) = \phi'(0) = \langle \nabla f(a), d\rangle.$$

But we know that $f'(a; d)$ is not always equal to $\langle \nabla f(a), d\rangle$. This means that the Chain Rule is not universally true and must involve conditions on the function f. Clearly, unless the function f is Gâteaux-differentiable, the chain rule cannot hold. For an in-depth treatment of this matter, consult Ortega and Rheinboldt [174].

8.8 Exercises

Ex. 8.1 *Let Q be a real, positive-definite symmetric matrix. Define the Q-inner product on $\mathbb{R}^J$ to be*

$$\langle x, y\rangle_Q = x^T Qy = \langle x, Qy\rangle,$$

and the Q-norm to be

$$||x||_Q = \sqrt{\langle x, x\rangle_Q}.$$

Show that, if $\nabla f(a)$ is the Fréchet derivative of $f(x)$ at $x = a$, for the usual Euclidean norm, then $Q^{-1}\nabla f(a)$ is the Fréchet derivative of $f(x)$ at $x = a$, for the Q-norm. Hint: Use the inequality

$$\sqrt{\lambda_J}||h||_2 \leq ||h||_Q \leq \sqrt{\lambda_1}||h||_2,$$

where λ_1 and λ_J denote the greatest and smallest eigenvalues of Q, respectively.

Ex. 8.2 [23, Ex. 10, p. 134] *For (x, y) not equal to $(0, 0)$, let*

$$f(x, y) = \frac{x^a y^b}{x^p + y^q},$$

with $f(0, 0) = 0$. In each of the five cases below, determine if the function is continuous, Gâteaux-, Fréchet-, or continuously differentiable at $(0, 0)$.

(a) $a = 2$, $b = 3$, $p = 2$, *and* $q = 4$;

(b) $a = 1$, $b = 3$, $p = 2$, *and* $q = 4$;

(c) $a = 2$, $b = 4$, $p = 4$, *and* $q = 8$;

(d) $a = 1$, $b = 2$, $p = 2$, *and* $q = 2$;

(e) $a = 1$, $b = 2$, $p = 2$, *and* $q = 4$.

Chapter 9

Convex Functions

9.1 Chapter Summary

In this chapter we investigate further the properties of convex functions of one and several variables, in preparation for our discussion of convex programming and iterative optimization algorithms.

9.2 Functions of a Single Real Variable

We begin by recalling some of the basic results concerning functions of a single real variable.

9.2.1 Fundamental Theorems

- The Intermediate Value Theorem (IVT):

  Theorem 9.1 *Let $f(x)$ be continuous on the interval $[a,b]$. If d is between $f(a)$ and $f(b)$, then there is c between a and b with $f(c) = d$.*

- Rolle's Theorem:

  Theorem 9.2 *Let $f(x)$ be continuous on the closed interval $[a,b]$ and differentiable on (a,b), with $f(a) = f(b)$. Then, there is c in (a,b) with $f'(c) = 0$.*

- The Mean Value Theorem (MVT):

  Theorem 9.3 *Let $f(x)$ be continuous on the closed interval $[a,b]$ and differentiable on (a,b). Then, there is c in (a,b) with*

$$f(b) - f(a) = f'(c)(b - a).$$

- A MVT for Integrals:

  Theorem 9.4 *Let $g(x)$ be continuous and $h(x)$ integrable with constant sign on the interval $[a,b]$. Then there is c in (a,b) such that*

$$\int_a^b g(x)h(x)dx = g(c)\int_a^b h(x)dx.$$

- The Extended Mean Value Theorem (EMVT):

  Theorem 9.5 *Let $f(x)$ be twice differentiable on the interval (u,v) and let a and b be in (u,v). Then there is c between a and b with*

$$f(b) = f(a) + f'(a)(b - a) + \frac{1}{2}f''(c)(b - a)^2.$$

If $f(x)$ is a function with $f''(x) > 0$ for all x and $f'(a) = 0$, then, from the EMVT, we know that $f(b) > f(a)$, unless $b = a$, so that $x = a$ is a global minimizer of the function $f(x)$. As we shall see, such functions are strictly convex.

9.2.2 Proof of Rolle's Theorem

The IVT is a direct consequence of the completeness of $\mathbb{R}$. To prove Rolle's Theorem, we simply note that either f is constant, in which case $f'(x) = 0$ for all x in (a, b), or it has a local maximum or minimum at c in (a, b), in which case $f'(c) = 0$.

9.2.3 Proof of the Mean Value Theorem

The main use of Rolle's Theorem is to prove the Mean Value Theorem. Let

$$g(x) = f(x) - \Big(\frac{f(b) - f(a)}{b - a}\Big)(x - a).$$

Then $g(a) = g(b)$ and so there is $c \in (a, b)$ with $g'(c) = 0$, or

$$f(b) - f(a) = f'(c)(b - a).$$

9.2.4 A Proof of the MVT for Integrals

We now prove the Mean Value Theorem for Integrals. Since $g(x)$ is continuous on the interval $[a, b]$, it takes on its minimum value, say m, and its maximum value, say M, and, by the Intermediate Value Theorem, $g(x)$ also takes on any value in the interval $[m, M]$. Assume, without loss of generality, that $h(x) \geq 0$, for all x in the interval $[a, b]$, so that $\int_a^b h(x)dx \geq 0$. Then we have

$$m \int_a^b h(x)dx \leq \int_a^b g(x)h(x)dx \leq M \int_a^b h(x)dx,$$

which says that the ratio

$$\frac{\int_a^b g(x)h(x)dx}{\int_a^b h(x)dx}$$

lies in the interval $[m, M]$. Consequently, there is a value c in (a, b) for which $g(c)$ has the value of this ratio. This completes the proof.

9.2.5 Two Proofs of the EMVT

Now we present two proofs of the EMVT. We begin by using integration by parts, with $u(x) = f'(x)$ and $v(x) = x - b$, to get

$$f(b) - f(a) = \int_a^b f'(x)dx = f'(x)(x - b)|_a^b - \int_a^b f''(x)(x - b)dx,$$

or

$$f(b) - f(a) = -f'(a)(a - b) - \int_a^b f''(x)(x - b)dx.$$

Then, using the MVT for integrals, with $g(x) = f''(x)$ assumed to be continuous, and $h(x) = x - b$, we have

$$f(b) = f(a) + f'(a)(b - a) - f''(c)\int_a^b (x - b)dx,$$

from which the assertion of the theorem follows immediately.

A second proof of the EMVT, which does not require that $f''(x)$ be continuous, is as follows. Let a and b be fixed and set

$$F(x) = f(x) + f'(x)(b - x) + A(b - x)^2,$$

for some constant A to be determined. Then $F(b) = f(b)$. Select A so that $F(a) = f(b)$. Then $F(b) = F(a)$, so there is c in (a, b) with $F'(c) = 0$, by the MVT, or, more simply, from Rolle's Theorem. Therefore,

$$0 = F'(c) = f'(c)+f''(c)(b-c)+f'(c)(-1)-2A(b-c) = (f''(c)-2A)(b-c).$$

So $A = \frac{1}{2}f''(c)$ and

$$F(x) = f(x) + f'(x)(b - x) + \frac{1}{2}f''(c)(b - x)^2,$$

from which we get

$$F(a) = f(b) = f(a) + f'(a)(b - a) + \frac{1}{2}f''(c)(b - a)^2.$$

This completes the second proof. ∎

9.2.6 Lipschitz Continuity

Let $f : \mathbb{R} \to \mathbb{R}$ be a differentiable function. From the Mean-Value Theorem we know that

$$f(b) = f(a) + f'(c)(b - a),$$

for some c between a and b. If there is a constant L with $|f'(x)| \leq L$ for all x, that is, the derivative is bounded, then we have

$$|f(b) - f(a)| \leq L|b - a|, \tag{9.1}$$

for all a and b; functions that satisfy Equation (9.1) are said to be *L-Lipschitz continuous*.

9.2.7 The Convex Case

We focus now on the special case of convex functions. Earlier, we said that a proper function $g : \mathbb{R} \rightarrow (-\infty, \infty]$ is convex if its epigraph is a convex set, in which case the effective domain of the function g must be a convex set, since it is the orthogonal projection of the convex epigraph. For a real-valued function g defined on the whole real line we have several conditions on g that are equivalent to being a convex function.

Proposition 9.1 *Let $g : \mathbb{R} \rightarrow \mathbb{R}$. The following are equivalent:*

(1) the epigraph of $g(x)$ is convex;

(2) for all points $a < x < b$ in $\mathbb{R}$

$$g(x) \leq \frac{g(b) - g(a)}{b - a}(x - a) + g(a);$$

(3) for all points $a < x < b$ in $\mathbb{R}$

$$g(x) \leq \frac{g(b) - g(a)}{b - a}(x - b) + g(b);$$

(4) for all points a and b in $\mathbb{R}$ and for all α in the interval $(0, 1)$

$$g((1 - \alpha)a + \alpha b) \leq (1 - \alpha)g(a) + \alpha g(b).$$

The proof of Proposition 9.1 is left as an exercise.

As a result of Proposition 9.1, we can use the following definition of a convex real-valued function.

Definition 9.1 *A function $g : \mathbb{R} \rightarrow \mathbb{R}$ is called* convex *if, for each pair of distinct real numbers a and b, the line segment connecting the two points $A = (a, g(a))$ and $B = (b, g(b))$ is on or above the graph of $g(x)$; that is, for every α in $(0, 1)$,*

$$g((1 - \alpha)a + \alpha b) \leq (1 - \alpha)g(a) + \alpha g(b).$$

If the inequality is always strict, then $g(x)$ is strictly convex.

The function $g(x) = x^2$ is a simple example of a convex function. If $g(x)$ is convex, then $g(x)$ is continuous, as well [176, p. 47]. It follows from Proposition 9.1 that, if $g(x)$ is convex, then, for every triple of points $a < x < b$, we have

$$\frac{g(x) - g(a)}{x - a} \leq \frac{g(b) - g(a)}{b - a} \leq \frac{g(b) - g(x)}{b - x}.$$

Therefore, for fixed a, the ratio

$$\frac{g(x) - g(a)}{x - a}$$

is an increasing function of x, and, for fixed b, the ratio

$$\frac{g(b) - g(x)}{b - x}$$

is an increasing function of x.

If we allow g to take on the value $+\infty$, then we say that g is convex if and only if, for all points a and b in $\mathbb{R}$ and for all α in the interval $(0, 1)$,

$$g((1 - \alpha)a + \alpha b) \leq (1 - \alpha)g(a) + \alpha g(b).$$

If $g(x)$ is a differentiable function, then convexity can be expressed in terms of properties of the derivative, $g'(x)$; for every triple of points $a < x < b$, we have

$$g'(a) \leq \frac{g(b) - g(a)}{b - a} \leq g'(b).$$

If $g(x)$ is differentiable and convex, then $g'(x)$ is an increasing function. In fact, the converse is also true, as we shall see shortly.

Recall that the line tangent to the graph of $g(x)$ at the point $x = a$ has the equation

$$y = g'(a)(x - a) + g(a).$$

Theorem 9.6 *For the differentiable function $g(x)$, the following are equivalent:*

(1) $g(x)$ is convex;

(2) for all a and x we have

$$g(x) \geq g(a) + g'(a)(x - a); \tag{9.2}$$

(3) the derivative, $g'(x)$, is an increasing function, or, equivalently,

$$(g'(x) - g'(a))(x - a) \geq 0,$$

for all a and x.

Proof: Assume that $g(x)$ is convex. If $x > a$, then

$$g'(a) \leq \frac{g(x) - g(a)}{x - a},$$

while, if $x < a$, then

$$\frac{g(a) - g(x)}{a - x} \leq g'(a).$$

In either case, the inequality in (9.2) holds. Now, assume that the inequality in (9.2) holds. Then

$$g(x) \geq g'(a)(x - a) + g(a),$$

and

$$g(a) \geq g'(x)(a - x) + g(x).$$

Adding the two inequalities, we obtain

$$g(a) + g(x) \geq (g'(x) - g'(a))(a - x) + g(a) + g(x),$$

from which we conclude that

$$(g'(x) - g'(a))(x - a) \geq 0. \tag{9.3}$$

So $g'(x)$ is increasing. Finally, we assume the derivative is increasing and show that $g(x)$ is convex. If $g(x)$ is not convex, then there are points $a < b$ such that, for all x in (a, b),

$$\frac{g(x) - g(a)}{x - a} > \frac{g(b) - g(a)}{b - a}.$$

By the Mean Value Theorem there is c in (a, b) with

$$g'(c) = \frac{g(b) - g(a)}{b - a}.$$

Select x in the interval (a, c). Then there is d in (a, x) with

$$g'(d) = \frac{g(x) - g(a)}{x - a}.$$

Then $g'(d) > g'(c)$, which contradicts the assumption that $g'(x)$ is increasing. This concludes the proof. ∎

If $g(x)$ is twice differentiable, we can say more. If we multiply both sides of the inequality in (9.3) by $(x - a)^{-2}$, we find that

$$\frac{g'(x) - g'(a)}{x - a} \geq 0, \tag{9.4}$$

for all x and a. This inequality suggests the following theorem.

Theorem 9.7 *If $g(x)$ is twice differentiable, then $g(x)$ is convex if and only if $g''(x) \geq 0$, for all x.*

Proof: According to the Mean Value Theorem, as applied to the function $g'(x)$, for any points $a < b$ there is c in (a, b) with $g'(b) - g'(a) = g''(c)(b-a)$. If $g''(x) \geq 0$, the right side of this equation is nonnegative, so the left side is also. Now assume that $g(x)$ is convex, which implies that $g'(x)$ is an increasing function. Since $g'(x+h) - g'(x) \geq 0$ for all $h > 0$, it follows that $g''(x) \geq 0$. ∎

The following result, as well as its extension to higher dimensions, will be helpful in our study of iterative optimization.

Theorem 9.8 *Let $h(x)$ be convex and differentiable and its derivative, $h'(x)$, nonexpansive, that is,*

$$|h'(b) - h'(a)| \leq |b - a|,$$

for all a and b. Then $h'(x)$ is firmly nonexpansive, which means that

$$(h'(b) - h'(a))(b - a) \geq (h'(b) - h'(a))^2.$$

Proof: Assume that $h'(b) - h'(a) \neq 0$, since the alternative case is trivial. If $h'(x)$ is nonexpansive, then the inequality in (9.4) tells us that

$$0 \leq \frac{h'(b) - h'(a)}{b - a} \leq 1,$$

so that

$$\frac{b - a}{h'(b) - h'(a)} \geq 1.$$

Now multiply both sides by $(h'(b) - h'(a))^2$. ∎

In the next section we extend these results to functions of several variables.

9.3 Functions of Several Real Variables

In this section we consider the continuity and differentiability of a function of several variables. For more details, see the chapter on differentiability.

9.3.1 Continuity

In addition to real-valued functions $f : \mathbb{R}^J \to \mathbb{R}$, we shall also be interested in vector-valued functions $F : \mathbb{R}^J \to \mathbb{R}^I$, such as $F(x) = \nabla f(x)$, whose range is in $\mathbb{R}^J$, not in $\mathbb{R}$.

Definition 9.2 *We say that* $F : \mathbb{R}^J \to \mathbb{R}^I$ *is* continuous *at* $x = a$ *if*

$$\lim_{x \to a} f(x) = f(a);$$

that is, $\|f(x) - f(a)\|_2 \to 0$, *as* $\|x - a\|_2 \to 0$.

Definition 9.3 *We say that* $F : \mathbb{R}^J \to \mathbb{R}^I$ *is* L-Lipschitz, *or an* L-Lipschitz continuous function, *with respect to the 2-norm, if there is* $L > 0$ *such that*

$$\|F(b) - F(a)\|_2 \leq L\|b - a\|_2,$$

for all a *and* b *in* $\mathbb{R}^J$.

9.3.2 Differentiability

Let $F : D \subseteq \mathbb{R}^J \to \mathbb{R}^I$ be a $\mathbb{R}^I$-valued function of J real variables, defined on domain D with nonempty interior $\text{int}(D)$.

Definition 9.4 *The function* $F(x)$ *is said to be* Fréchet-differentiable, *or just* differentiable, *at point* x^0 *in int*(D) *if there is an* I *by* J *matrix* $F'(x^0)$ *such that*

$$\lim_{h \to 0} \frac{1}{||h||_2}[F(x^0 + h) - F(x^0) - F'(x^0)h] = 0.$$

It can be shown that, if F is differentiable at $x = x^0$, then F is continuous there as well [114].

If $f : \mathbb{R}^J \to \mathbb{R}$ is differentiable, then $f'(x^0) = \nabla f(x^0)$, the gradient of f at x^0. The function $f(x)$ is differentiable if each of its first partial derivatives is continuous. If f is finite and convex and differentiable on an open convex set C, then ∇f is continuous on C [182, Corollary 25.5.1].

If the derivative $f' : \mathbb{R}^J \to \mathbb{R}^J$ is, itself, differentiable, then $f'' : \mathbb{R}^J \to \mathbb{R}^{J \times J}$, and $f''(x) = H(x) = \nabla^2 f(x)$, the Hessian matrix whose entries are the second partial derivatives of f. The function $f(x)$ will be twice differentiable if each of the second partial derivatives is continuous. In that case, the mixed second partial derivatives are independent of the order of the variables, the Hessian matrix is symmetric, and the chain rule applies.

Let $f : \mathbb{R}^J \to \mathbb{R}$ be a differentiable function. The Mean-Value Theorem for f is the following.

Theorem 9.9 (The Mean Value Theorem) *For any two points* a *and* b *in* $\mathbb{R}^J$, *there is* α *in* $(0, 1)$ *such that*

$$f(b) = f(a) + \langle \nabla f((1 - \alpha)a + \alpha b), b - a \rangle. \tag{9.5}$$

Proof: To prove this, we parameterize the line segment between the points a and b as $x(t) = a + t(b - a)$. Then we define $g(t) = f(x(t))$. We can apply the ordinary mean value theorem to $g(t)$, to get

$$g(1) = g(0) + g'(\alpha),$$

for some α in the interval $[0, 1]$. The derivative of $g(t)$ is

$$g'(t) = \langle \nabla f(x(t)), b - a \rangle,$$

where

$$\nabla f(x(t)) = \Big(\frac{\partial f}{\partial x_1}(x(t)), ..., \frac{\partial f}{\partial x_J}(x(t)) \Big)^T.$$

Therefore,

$$g'(\alpha) = \langle \nabla f(x(\alpha), b - a \rangle.$$

Since $x(\alpha) = (1 - \alpha)a + \alpha b$, the proof is complete. ∎

If there is a constant L with $||\nabla f(x)||_2 \leq L$ for all x, that is, the gradient is bounded in norm, then we have

$$|f(b) - f(a)| \leq L||b - a||_2,$$

for all a and b; such functions are then L-Lipschitz continuous. We can study multivariate functions $f : \mathbb{R}^J \to \mathbb{R}$ by using them to construct functions of a single real variable, given by

$$\phi(t) = f(x^0 + t(x - x^0)),$$

where x and x^0 are fixed (column) vectors in $\mathbb{R}^J$. If $f(x)$ is differentiable, then

$$\phi'(t) = \langle \nabla f(x^0 + t(x - x^0)), x - x^0 \rangle.$$

If $f(x)$ is twice continuously differentiable, then

$$\phi''(t) = (x - x^0)^T \nabla^2 f(x^0 + t(x - x^0))(x - x^0).$$

Definition 9.5 *A function* $f : \mathbb{R}^J \to \mathbb{R}$ *is called* coercive *if*

$$\lim_{||x||_2 \to +\infty} f(x) = +\infty.$$

Definition 9.6 *A function* $f : \mathbb{R}^J \to \mathbb{R}$ *is called* super-coercive *if*

$$\lim_{||x||_2 \to +\infty} \frac{f(x)}{||x||_2} = +\infty.$$

We have the following proposition, whose proof is left as Exercise 9.3.

Proposition 9.2 *Let $f : \mathbb{R}^J \to \mathbb{R}$ be a super-coercive differentiable function. Then the gradient operator $\nabla f : \mathbb{R}^J \to \mathbb{R}^J$ is onto $\mathbb{R}^J$; that is, for every $y \in \mathbb{R}^J$ there is $x \in \mathbb{R}^J$ with $\nabla f(x) = y$.*

For example, the function $f : \mathbb{R} \to \mathbb{R}$ given by $f(x) = \frac{1}{2}x^2$ satisfies the conditions of the proposition and its derivative is $f'(x) = x$, whose range is all of $\mathbb{R}$. In contrast, the function $g(x) = \frac{1}{3}x^3$ is not coercive and its derivative, $g'(x) = x^2$, does not have all of $\mathbb{R}$ for its range.

9.3.3 Second Differentiability

We assume, throughout this subsection, that $f : \mathbb{R}^J \to \mathbb{R}$ has continuous second partial derivatives. Then $H(x) = \nabla^2 f(x)$, the Hessian matrix of f at the point x, has for its entries the second partial derivatives of f at x, and is symmetric. The following theorems are fundamental in describing local maxima and minima of f.

Theorem 9.10 *Let x and x^* be points in $\mathbb{R}^J$. Then there is a point z on the line segment $[x^*, x]$ connecting x with x^* such that*

$$f(x) = f(x^*) + \nabla f(x^*) \cdot (x - x^*) + \frac{1}{2}(x - x^*) \cdot H(z)(x - x^*). \tag{9.6}$$

Consider the problem of optimizing the function $f(x)$. The first step is to find the critical points. Assume that f is twice differentiable and that x^* is a critical point, so that $\nabla f(x^*) = 0$. Then, from Equation (9.6) we have

$$f(x) = f(x^*) + \frac{1}{2}(x - x^*) \cdot H(z)(x - x^*),$$

so that

$$f(x) - f(x^*) = \frac{1}{2}(x - x^*) \cdot H(z)(x - x^*).$$

The behavior of the function f around x^* depends on the quadratic form $(x - x^*) \cdot H(z)(x - x^*)$. We have the following theorem.

Theorem 9.11 *Let x^* be a critical point, that is, $\nabla f(x^*) = 0$. Then*

(1) x^* is a global minimizer of $f(x)$ if $(x - x^*) \cdot H(z)(x - x^*) \geq 0$ for all x and for all z in $[x^*, x]$;

(2) x^* is a strict global minimizer of $f(x)$ if $(x - x^*) \cdot H(z)(x - x^*) > 0$ for all $x \neq x^*$ and for all z in $[x^*, x]$;

(3) x^* is a global maximizer of $f(x)$ if $(x - x^*) \cdot H(z)(x - x^*) \leq 0$ for all x and for all z in $[x^*, x]$;

(4) x^* is a strict global maximizer of $f(x)$ if $(x - x^*) \cdot H(z)(x - x^*) < 0$ for all $x \neq x^*$ and for all z in $[x^*, x]$.

9.3.4 Finding Maxima and Minima

Assume that $g : \mathbb{R}^J \to \mathbb{R}$ is differentiable and attains its minimum value. We want to minimize the function $g(x)$. Solving $\nabla g(x) = 0$ to find the optimal $x = x^*$ may not be easy, so we may turn to an iterative algorithm for finding roots of $\nabla g(x)$, or one that minimizes $g(x)$ directly. In the latter case, we may again consider a steepest descent algorithm of the form

$$x^{k+1} = x^k - \gamma \nabla g(x^k),$$

for some $\gamma > 0$. We denote by T the operator

$$Tx = x - \gamma \nabla g(x).$$

Then, using $\nabla g(x^*) = 0$, we find that

$$||x^* - x^{k+1}||_2 = ||Tx^* - Tx^k||_2.$$

We would like to know if there are choices for γ that imply convergence of the iterative sequence. As in the case of functions of a single variable, for functions $g(x)$ that are *convex*, the answer is yes.

9.3.5 Solving $F(x) = 0$ through Optimization

Consider a function $f(x) : \mathbb{R}^J \to \mathbb{R}$ that is strictly convex and has a unique global minimum at $\hat{x}$. If $F(x) = \nabla f(x)$ for all x, then $F(\hat{x}) = 0$. In some cases it may be simpler to minimize the function $f(x)$ than to solve for a zero of $F(x)$.

If $F(x)$ is not a gradient of any function $f(x)$, we may still be able to find a zero of $F(x)$ by minimizing some function. For example, let $g(x) = \|x\|_2$. Then the function $f(x) = g(F(x))$ is minimized when $F(x) = 0$.

The function $F(x) = Ax - b$ need not have a zero. In such cases, we can minimize the function $f(x) = \frac{1}{2}\|Ax - b\|_2^2$ to obtain the least-squares solution, which then can be viewed as an approximate zero of $F(x)$.

9.3.6 When Is $F(x)$ a Gradient?

The following theorem is classical and extends the familiar "test for exactness" ; see Ortega and Rheinboldt [174].

Theorem 9.12 *Let $F : D \subseteq \mathbb{R}^J \to \mathbb{R}^J$ be continuously differentiable on an open convex set $D_0 \subseteq D$. Then there is a differentiable function $f : D_0 \to \mathbb{R}$ such that $F(x) = \nabla f(x)$ for all x in D_0 if and only if the derivative $F'(x)$ is symmetric, where $F'(x)$ is the J by J Jacobian matrix with entries*

$$(F'(x))_{mn} = \frac{\partial F_m(x)}{\partial x_n},$$

and

$$F(x) = (F_1(x), F_2(x), ..., F_J(x)).$$

Proof: If $F(x) = \nabla f(x)$ for all x in D_0 and is continuously differentiable, then the second partial derivatives of $f(x)$ are continuous, so that the mixed second partial derivatives of $f(x)$ are independent of the order of differentiation. In other words, the matrix $F'(x)$ is symmetric, where now $F'(x)$ is the Hessian matrix of $f(x)$.

For notational convenience, we present the proof of the converse only for the case of $J = 3$; the proof is the same in general. The proof in [174] is somewhat different.

Without loss of generality, we assume that the origin is a member of the set D_0. Define $f(x, y, z)$ by

$$f(x, y, z) = \int_0^x F_1(u, 0, 0)du + \int_0^y F_2(x, u, 0)du + \int_0^z F_3(x, y, u)du.$$

We prove that $\frac{\partial f}{\partial x}(x, y, z) = F_1(x, y, z)$.

The partial derivative of the first integral, with respect to x, is $F_1(x, 0, 0)$. The partial derivative of the second integral, with respect to x, obtained by differentiating under the integral sign, is

$$\int_0^y \frac{\partial F_2}{\partial x}(x, u, 0)du,$$

which, by the symmetry of the Jacobian matrix, is

$$\int_0^y \frac{\partial F_1}{\partial y}(x, u, 0)du = F_1(x, y, 0) - F_1(x, 0, 0).$$

The partial derivative of the third integral, with respect to x, obtained by differentiating under the integral sign, is

$$\int_0^z \frac{\partial F_3}{\partial x}(x, y, u)du,$$

which, by the symmetry of the Jacobian matrix, is

$$\int_0^z \frac{\partial F_1}{\partial z}(x, y, u)du = F_1(x, y, z) - F_1(x, y, 0).$$

We complete the proof by adding these three integral values. Similar calculations show that $\nabla f(x) = F(x)$. ∎

9.3.7 Lower Semi-Continuity

We begin with a definition.

Definition 9.7 *A proper function f from $\mathbb{R}^J$ to $(-\infty, \infty]$ is* lower semi-continuous *if $f(x) = \liminf f(y)$, as $y \to x$.*

The following theorem shows the importance of lower semi-continuity.

Theorem 9.13 [182, **Theorem 7.1**] *Let f be an arbitrary proper function from $\mathbb{R}^J$ to $(-\infty, \infty]$. Then the following conditions are equivalent:*

(1) f is lower semi-continuous throughout $\mathbb{R}^J$;

(2) for every real α, the set $\{x|f(x) \leq \alpha\}$ is closed;

(3) the epigraph of $f(x)$ is closed.

As an example, consider the function $f(x)$ defined for $-1 \leq x < 0$ by $f(x) = -x - 1$, and for $0 < x \leq 1$ by $f(x) = -x + 1$. If we define $f(0) = -1$, then $f(x)$ becomes lower semi-continuous at $x = 0$ and the epigraph becomes closed. If we define $f(0) = 1$, the function is upper semi-continuous at $x = 0$, but is no longer lower semi-continuous there; its epigraph is no longer closed.

It is helpful to recall the following theorem:

Theorem 9.14 *Let $f : \mathbb{R}^J \to \mathbb{R}$ be LSC and let $C \subseteq \mathbb{R}^J$ be nonempty, closed, and bounded. Then there is a in C with $f(a) \leq f(x)$, for all x in C.*

9.3.8 The Convex Case

We begin with some definitions.

Definition 9.8 *The proper function $g(x) : \mathbb{R}^J \to (-\infty, \infty]$ is said to be* convex *if, for each pair of distinct vectors a and b and for every α in the interval $(0, 1)$ we have*

$$g((1-\alpha)a + \alpha b) \leq (1-\alpha)g(a) + \alpha g(b).$$

If the inequality is always strict, then $g(x)$ is called strictly convex.

The function $g(x)$ is convex if and only if, for every x and z in $\mathbb{R}^J$ and real t, the function $f(t) = g(x + tz)$ is a convex function of t. Therefore, the theorems for the multi-variable case can also be obtained from previous results for the single-variable case.

Definition 9.9 *A proper convex function g is* closed *if it is lower semi-continuous.*

A proper convex function g is closed if and only if its epigraph is a closed set.

Definition 9.10 *The* closure *of a proper convex function g is the function* cl$g$ *defined by*

$$\mathrm{cl}g(x) = \liminf_{y \to x} g(y).$$

The function cl$g$ is convex and lower semi-continuous and agrees with g, except perhaps at points of the relative boundary of dom(g). The epigraph of cl$g$ is the closure of the epigraph of g.

If g is convex and finite on an open subset of dom(g), then g is continuous there, as well [182]. In particular, we have the following theorem.

Theorem 9.15 *Let $g : \mathbb{R}^J \to \mathbb{R}$ be convex and finite-valued on $\mathbb{R}^J$. Then g is continuous.*

Let $\iota_C(x)$ be the *indicator function* of the closed convex set C, that is, $\iota_C(x) = 0$ if $x \in C$, and $\iota_C(x) = +\infty$, otherwise. This function is lower semi-continuous, convex, but not continuous at points on the boundary of C. If we had defined $\iota_C(x)$ to be, say, 1, for x not in C, then the function would have been lower semi-continuous, and finite everywhere, but would no longer be convex.

As in the case of functions of a single real variable, we have several equivalent notions of convexity for differentiable functions of more than one variable.

Theorem 9.16 *Let $g : \mathbb{R}^J \to \mathbb{R}$ be differentiable. The following are equivalent:*

(1) $g(x)$ is convex;

(2) for all x and y we have

$$g(x) \geq g(y) + \langle \nabla g(y), x - y \rangle\,;$$

(3) for all x and y we have

$$\langle \nabla g(x) - \nabla g(y), x - y \rangle \geq 0.$$

Proof: First, we show that (1) implies (2). According to the Mean Value Theorem, Equation (9.5), we have

$$g(x) = g(y) + \langle \nabla g(\alpha x + (1 - \alpha)y), x - y \rangle,$$

for some α in the interval $(0, 1)$. From the definition of the gradient, we know that

$$\lim_{\alpha \downarrow 0} \frac{1}{\alpha}(g(y + \alpha(x - y)) - g(y) - \langle \nabla g(y), \alpha(x - y)\rangle) = 0.$$

Using (1), we have

$$g(x) - g(y) - \langle \nabla g(y), x - y\rangle \geq \frac{1}{\alpha}(g(y + \alpha(x - y)) - g(y) - \langle \nabla g(y), \alpha(x - y)\rangle),$$

for all α. Then we take limits on both sides of this inequality, as $\alpha \downarrow 0$.

Next we show that (2) implies (1). We have

$$g(x) - g((1 - \alpha)x + \alpha y) \geq \alpha\langle \nabla g((1 - \alpha)x + \alpha y), x - y\rangle,$$

and

$$g(y) - g((1 - \alpha)x + \alpha y) \geq -(1 - \alpha)\langle \nabla g((1 - \alpha)x + \alpha y), x - y\rangle.$$

Therefore,

$$(1 - \alpha)g(x) - (1 - \alpha)g((1 - \alpha)x + \alpha y) \geq (1 - \alpha)\alpha\langle \nabla g((1 - \alpha)x + \alpha y), x - y\rangle,$$

and

$$\alpha g(y) - \alpha g((1 - \alpha)x + \alpha y) \geq -\alpha(1 - \alpha)\langle \nabla g((1 - \alpha)x + \alpha y), x - y\rangle.$$

Now add the last two inequalities.

Showing that (2) implies (3) is easy, so we conclude the proof by showing that (3) implies (2). Once again, the Mean Value Theorem tells us that

$$g(x) = g(y) + \langle \nabla g(\alpha x + (1 - \alpha)y), x - y\rangle,$$

for some α in the interval $(0, 1)$. Using (3) we have

$$\langle \nabla g(\alpha x + (1 - \alpha)y) - \nabla g(y), \alpha x + (1 - \alpha)y - y\rangle \geq 0,$$

for all α. Therefore,

$$\langle \nabla g(\alpha x + (1 - \alpha)y) - \nabla g(y), x - y\rangle \geq 0,$$

for all α. This completes the proof. ∎

Corollary 9.1 *The function* $g(x) = \frac{1}{2}\left(\|x\|_2^2 - \|x - P_C x\|_2^2\right)$ *is convex.*

Proof: We show later in Corollary 12.1 that the gradient of $g(x)$ is $\nabla g(x) = P_C x$. From the inequality (4.4) we know that

$$\langle P_C x - P_C y, x - y\rangle \geq 0,$$

for all x and y. Therefore, $g(x)$ is convex, by Theorem 9.16. ∎

Definition 9.11 *Let* $g : \mathbb{R}^J \to \mathbb{R}$ *be convex and differentiable. Then the* Bregman distance, *from* x *to* y, *associated with* g *is*

$$D_g(x, y) = g(x) - g(y) - \langle \nabla g(y), x - y \rangle.$$

Since g is convex, Theorem 9.16 tells us that $D_g(x, y) \geq 0$, for all x and y. Also, for each fixed y, the function $d(x) = D_g(x, y)$ is $g(x)$ plus a linear function of x; therefore, $d(x)$ is also convex.

If we impose additional restrictions on g, then we can endow $D_g(x, y)$ with additional properties usually associated with a distance measure; for example, if g is strictly convex, then $D_g(x, y) = 0$ if and only if $x = y$.

As in the case of functions of a single variable, we can say more when the function $g(x)$ is twice differentiable. To guarantee that the second derivative matrix is symmetric, we assume that the second partial derivatives are continuous. Note that, by the chain rule again, $f''(t) = z^T \nabla^2 g(x + tz) z$.

Theorem 9.17 *Let each of the second partial derivatives of* $g(x)$ *be continuous, so that* $g(x)$ *is twice continuously differentiable. Then* $g(x)$ *is convex if and only if the second derivative matrix* $\nabla^2 g(x)$ *is nonnegative definite, for each* x.

9.4 Sub-Differentials and Sub-Gradients

The following proposition describes the relationship between hyperplanes supporting the epigraph of a differentiable function and its gradient. The proof is left as Exercise 9.5.

Proposition 9.3 *Let* $g : \mathbb{R}^J \to \mathbb{R}$ *be a convex function that is differentiable at the point* x^0. *Then there is a unique hyperplane* H *supporting the epigraph of* g *at the point* $(x^0, g(x^0))$ *and* H *can be written as*

$$H = \{z \in \mathbb{R}^{J+1} | \langle a, z \rangle = \gamma\},$$

for

$$a^T = (\nabla g(x^0)^T, -1)$$

and

$$\gamma = \langle \nabla g(x^0), x^0 \rangle - g(x^0).$$

Now we want to extend Proposition 9.3 to the case of nondifferentiable functions. Suppose that $g : \mathbb{R}^J \to (-\infty, +\infty]$ is convex and $g(x)$ is finite for x in the nonempty convex set C. If x^0 is in the interior of C, then g is continuous at x^0. Applying the Support Theorem to the epigraph of cl$g$, we obtain the following theorem.

Theorem 9.18 *If x^0 is an interior point of the set C, then there is a nonzero vector u with*

$$g(x) \geq g(x^0) + \langle u, x - x^0 \rangle, \tag{9.7}$$

for all x.

Proof: The point $(x^0, g(x^0))$ is a boundary point of the epigraph of g. According to the Support Theorem, there is a nonzero vector $a = (b, c)$ in $\mathbb{R}^{J+1}$, with b in $\mathbb{R}^J$ and c real, such that

$$\langle b, x \rangle + cr = \langle a, (x, r) \rangle \leq \langle a, (x^0, g(x^0)) \rangle = \langle b, x^0 \rangle + cg(x^0),$$

for all (x, r) in the epigraph of g, that is, all (x, r) with $g(x) \leq r$. The real number c cannot be positive, since $\langle b, x \rangle + cr$ is bounded above, while r can be increased arbitrarily. Also c cannot be zero; if $c = 0$, then b cannot be zero and we would have $\langle b, x \rangle \leq \langle b, x^0 \rangle$ for all x in C. But, since x^0 is in the interior of C, there is $t > 0$ such that $x = x^0 + tb$ is in C. So $c < 0$. We then select $u = -\frac{1}{c}b$. The inequality in (9.7) follows. ∎

Note that it can happen that $b = 0$; therefore $u = 0$ is possible; see Exercise 9.12.

Definition 9.12 *A vector u is said to be a* sub-gradient *of the function $g(x)$ at $x = x^0$ if, for all x, we have*

$$g(x) \geq g(x^0) + \langle u, x - x^0 \rangle.$$

The collection of all sub-gradients of g at $x = x^0$ is called the sub-differential *of g at $x = x^0$, denoted $\partial g(x^0)$. The* domain *of ∂g is the set* dom $\partial g = \{x | \partial g(x) \neq \emptyset\}$.

As an example, consider the function $f(x) = x^2$. The epigraph of $f(x)$ is the set of all points in the x, y-plane on or above the graph of $f(x)$. At the point $(1, 1)$ on the boundary of the epigraph the supporting hyperplane is just the tangent line, which can be written as $y = 2x - 1$ or $2x - y = 1$. The outward normal vector is $a = (b, c) = (2, -1)$. Then $u = b = 2 = f'(1)$.

As we have seen, if $f : \mathbb{R}^J \to \mathbb{R}$ is differentiable, then an outward normal vector to the hyperplane supporting the epigraph at the boundary point $(x_0, f(x_0))$ is the vector

$$a = (b^T, c^T)^T = (\nabla f(x_0)^T, -1)^T.$$

So $b = u = \nabla f(x_0)$.

When $f(x)$ is not differentiable at $x = x_0$ there will be multiple hyperplanes supporting the epigraph of $f(x)$ at the boundary point $(x_0, f(x_0))$; the normals can be chosen to be $a = (b^T, -1)^T$, so that $b = u$ is a sub-gradient of $f(x)$ at $x = x_0$. For example, consider the function of real x

given by $g(x) = |x|$, and $x^0 = 0$. For any α with $|\alpha| \leq 1$, the graph of the straight line $y = \alpha x$ is a hyperplane supporting the epigraph of $g(x)$ at $x = 0$. Writing $\alpha x - y = 0$, we see that the vector $a = (b, c) = (\alpha, -1)$ is normal to the hyperplane. The constant $b = u = \alpha$ is a sub-gradient and for all x we have

$$g(x) = |x| \geq g(x^0) + \langle u, x - x^0 \rangle = \alpha x.$$

Let $g : \mathbb{R} \to \mathbb{R}$. Then m is in the sub-differential $\partial g(x_0)$ if and only if the line $y = mx + b$ passes through the point $(x_0, g(x_0))$ and $mx + b \leq g(x)$ for all x. As the reader is asked to show in Exercise 9.4, when g is differentiable at $x = x_0$ the only value of m that works is $m = g'(x_0)$, and the only line that works is the line tangent to the graph of g at $x = x_0$.

Theorem 9.18 says that the sub-differential of a convex function at an interior point of its domain is nonempty. If the sub-differential consists of a single vector, then g is differentiable at $x = x^0$ and that single vector is its gradient at $x = x^0$.

Note that, by the chain rule, $f'(t) = \nabla g(x + tz) \cdot z$, for the function $f(t) = g(x + tz)$.

Whenever $\nabla g(x)$ exists, it is the only sub-gradient for g at x. The following lemma, whose proof is left as Exercise 9.8, provides a further connection between the partial derivatives of g and the entries of any sub-gradient vector u.

Lemma 9.1 *Let $g : \mathbb{R}^J \to \mathbb{R}$ be a convex function, and u any sub-gradient of g at the point x. If $\frac{\partial g}{\partial x_j}(x)$ exists, then it is equal to u_j.*

Proof: In Exercise 9.8 the reader is asked to provide a proof. ∎

9.5 Sub-Gradients and Directional Derivatives

In this section we investigate the relationship between the sub-gradients of a convex function and its directional derivatives. Our discussion follows that of [23].

9.5.1 Some Definitions

Definition 9.13 *Let S be any subset of $\mathbb{R}^J$. A point x in S is said to be in the* core *of S, denoted* core(S), *if, for every vector z in $\mathbb{R}^J$, there is an $\epsilon > 0$, which may depend on z, such that, if $|t| \leq \epsilon$, then $x + tz$ and $x - tz$ are in S.*

The core of a set is a more general notion than the interior of a set; for x to be in the interior of S we must be able to find an $\epsilon > 0$ that works for all z. For example, let $S \subseteq \mathbb{R}^2$ be the set of all points on or above the graph of $y = x^2$, below or on the graph of $y = -x^2$ and the x-axis. The origin is then in the core of S, but is not in the interior of S. In Exercise 9.9 you will be asked to show that the core of S and the interior of S are the same, whenever S is convex.

Definition 9.14 *A function $f : \mathbb{R}^J \rightarrow (-\infty, +\infty]$ is* sub-linear *if, for all x and y in $\mathbb{R}^J$ and all nonnegative a and b,*

$$f(ax + by) \leq af(x) + bf(y).$$

We say that f is sub-additive *if*

$$f(x + y) \leq f(x) + f(y),$$

and positive homogeneous *if, for all positive λ,*

$$f(\lambda x) = \lambda f(x).$$

9.5.2 Sub-Linearity

We have the following proposition, the proof of which is left as Exercise 9.6.

Proposition 9.4 *A function $f : \mathbb{R}^J \rightarrow (-\infty, +\infty]$ is sub-linear if and only if it is both sub-additive and positive homogenous.*

Definition 9.15 *The* lineality space *of a sub-linear function f, denoted* lin(f), *is the largest subspace of $\mathbb{R}^J$ on which f is a linear functional.*

For example, take S to be a subspace of $\mathbb{R}^J$ and a a fixed member of $\mathbb{R}^J$. Define $f(x)$ by

$$f(x) = \langle a, P_S x\rangle \; + \|P_{S^\perp} x\|_2.$$

Then lin(f) is the subspace S.

Proposition 9.5 *Let $p : \mathbb{R}^J \rightarrow (-\infty, +\infty]$ be sub-linear and $S =$ lin(p). Then $p(s + x) = p(s) + p(x)$ for any s in S and any x in $\mathbb{R}^J$.*

Proof: We know that

$$p(s + x) \leq p(s) + p(x)$$

by the sub-additivity of p, so we need only show that

$$p(s+x) \geq p(s) + p(x).$$

Write

$$p(x) = p(x+s-s) \leq p(s+x) + p(-s) = p(s+x) - p(s),$$

so that

$$p(x) + p(s) \leq p(s+x).$$

∎

In this chapter, for notational convenience, we denote the one-sided directional derivative of f at x, in the direction of z, as

$$f'(x;z) = \lim_{t \downarrow 0} \frac{1}{t}(f(x+tz) - f(x)).$$

Proposition 9.6 *If $f : \mathbb{R}^J \to (-\infty, +\infty]$ is convex and x is in the core of* dom(f)*, then $f'(x;z)$, the directional derivative of f, at x and in the direction z, exists and is finite for all z and is a sub-linear function of z.*

Proof: For any z and real $t \neq 0$ let

$$g(z,t) = \frac{1}{t}(f(x+tz) - f(x)).$$

For $0 < t \leq s$ write

$$f(x+tz) = f\Big(\Big(1 - \frac{t}{s}\Big)x + \frac{t}{s}(x+sz)\Big) \leq \Big(1 - \frac{t}{s}\Big)f(x) + \frac{t}{s}f(x+sz).$$

It follows that

$$g(z,t) \leq g(z,s).$$

A similar argument gives

$$g(z,-s) \leq g(z,-t) \leq g(z,t) \leq g(z,s).$$

Since x lies in the core of dom(f), we can select $s > 0$ small enough so that both $g(z,-s)$ and $g(z,s)$ are finite. Therefore, as $t \downarrow 0$, the $g(z,t)$ are decreasing to the finite limit $f'(x;z)$; we have

$$-\infty < g(z,-s) \leq f'(x;z) \leq g(z,t) \leq g(z,s) < +\infty.$$

The sub-additivity of $f'(x;z)$ as a function of z follows easily from the inequality

$$g(z+y,t) \leq g(z,2t) + g(y,2t).$$

Proving the positive homogeneity of $f'(x;z)$ is easy. Therefore, $f'(x;z)$ is sub-linear in z. ∎

As pointed out by Borwein and Lewis in [23], the directional derivative of f is a local notion, defined only in terms of what happens to f near x, while the notion of a sub-gradient is clearly a global one. If f is differentiable at x, then we know that the derivative of f at x, which is then $\nabla f(x)$, can be used to express the directional derivatives of f at x:

$$f'(x;z) = \langle \nabla f(x), z \rangle.$$

We want to extend this relationship to sub-gradients of nondifferentiable functions.

9.5.3 Sub-Differentials and Directional Derivatives

We have the following proposition, whose proof is left as Exercise 9.7.

Proposition 9.7 *Let $f : \mathbb{R}^J \to (-\infty, +\infty]$ be convex and x in dom(f). Then u is a sub-gradient of f at x if and only if*

$$\langle u, z \rangle \leq f'(x;z)$$

for all z.

The main result of this subsection is the following theorem.

Theorem 9.19 *Let $f : \mathbb{R}^J \to (-\infty, +\infty]$ be convex and x in the core of dom(f). Let z be given. Then there is a u in $\partial f(x)$, with u depending on z, such that*

$$f'(x;z) = \langle u, z \rangle. \tag{9.8}$$

Therefore $f'(x;z)$ is the maximum of the quantities $\langle u, z \rangle$, as u ranges over the sub-differential $\partial f(x)$. In particular, the sub-differential is not empty.

Notice that Theorem 9.19 asserts that once z is selected, there will be a sub-gradient u for which Equation (9.8) holds. It does not assert that there will be one sub-gradient that works for all z; this happens only when there is only one sub-gradient, namely $\nabla f(x)$. The theorem also tells us that the function $f'(x;\cdot)$ is the support function of the closed convex set $C = \partial f(x)$.

We need the following proposition.

Proposition 9.8 *Let $p : \mathbb{R}^J \to (-\infty, +\infty]$ be sub-linear, and therefore convex, and let x lie in the core of dom(f). Define the function*

$$q(z) = p'(x;z).$$

Then $q(z)$ is sub-linear and has the following properties:

(1) $q(\lambda x) = \lambda p(x)$*, for all* λ*;*

(2) $q(z) \leq p(z)$*, for all* z*;*

(3) $\text{lin}(q)$ *contains the set* $\text{lin}(p) +$ *span*$\{x\}$.

Proof: If $t > 0$ is close enough to zero, then the quantity $1 + t\gamma$ is positive and

$$p(x + t\gamma x) = p((1 + t\gamma)x) = (1 + t\gamma)p(x),$$

by the positive homogeneity of p. Therefore,

$$q(\gamma x) = \lim_{t \downarrow 0} \frac{1}{t}\Big(p(x + t\gamma x) - p(x)\Big) = \gamma p(x).$$

Since

$$p(x + tz) \leq p(x) + tp(z),$$

we have

$$p(x + tz) - p(x) \leq tp(z),$$

from which $q(z) \leq p(z)$ follows immediately. Finally, let $\text{lin}(p) = S$. Then, by Proposition 9.5, we have

$$p(x + t(s + \gamma x)) = p(ts) + p((1 + t\gamma)x) = tp(s) + (1 + t\gamma)p(x),$$

for $t > 0$ close enough to zero. Therefore, we have

$$q(s + \gamma x) = p(s) + \gamma p(x).$$

From this it is easy to show that q is linear on $S+$ span$\{x\}$. ∎

Now we prove Theorem 9.19.

Proof of Theorem 9.19 Let y be fixed. Let $\{a^1, a^2, ..., a^J\}$ be a basis for $\mathbb{R}^J$, with $a^1 = y$. Let $p_0(z) = f'(x; z)$ and $p_1(z) = p_0'(a^1; z)$. Note that, since the function of z defined by $p_0(z) = f'(x; z)$ is convex and finite for all values of z, $p_0'(z; w)$ exists and is finite, for all z and all w. Therefore, $p_1(z) = p_0'(a^1; z)$ is sub-linear, and so convex, and finite for all z. The function $p_1(z)$ is linear on the span of the vector a^1. Because

$$p_0'(x; z) \leq p_0(x + z) - p_0(x)$$

and p_0 is sub-additive, we have

$$p_0'(x; z) = p_1(z) \leq p_0(z).$$

Continuing in this way, we define, for $k = 1, 2, ..., J$, $p_k(z) = p_{k-1}'(a^k; z)$. Then each $p_k(z)$ is sub-linear, and linear on the span of $\{a^1, ..., a^k\}$, and

$$p_k(z) \leq p_{k-1}(z).$$

Therefore, $p_J(z)$ is linear on all of $\mathbb{R}^J$. Finally, we have

$$p_J(y) \leq p_0(y) = p_0(a^1) = -p_0'(a^1; -a^1)$$

$$= -p_1(-a^1) = -p_1(-y) \leq -p_J(-y) = p_J(y),$$

with the last equality the result of the linearity of p_J. Therefore,

$$p_J(y) = f'(x; y).$$

Since $p_J(z)$ is a linear function, there is a vector u such that

$$p_J(z) = \langle u, z \rangle.$$

Since

$$p_J(z) = \langle u, z \rangle \leq f'(x; z) = p_0(z)$$

for all z, we know that $u \in \partial f(x)$. ∎

Theorem 9.19 shows that the sub-linear function $f'(x; \cdot)$ is the support functional for the set $\partial f(x)$. In fact, every lower semi-continuous sub-linear function is the support functional of some closed convex set, and every support functional of a closed convex set is a lower semi-continuous sub-linear function [129].

9.5.4 An Example

The function $f : \mathbb{R}^2 \to \mathbb{R}$ given by $f(x_1, x_2) = \frac{1}{2}x_1^2 + |x_2|$ has gradient $\nabla f(x_1, x_2) = (x_1, 1)^T$ if $x_2 > 0$, and $\nabla f(x_1, x_2) = (x_1, -1)^T$ if $x_2 < 0$, but is not differentiable when $x_2 = 0$. When $x_2 = 0$, the directional derivative function is

$$f'((x_1, 0); (z_1, z_2)) = x_1 z_1 + |z_2|,$$

and the sub-differential is

$$\partial f((x_1, 0)) = \{\phi = (x_1, \gamma)^T \,|\, -1 \leq \gamma \leq 1\}.$$

Therefore,

$$f'((x_1, 0); (z_1, z_2)) = \langle \phi, z \rangle,$$

with $\gamma = 1$ when $z_2 \geq 0$, and $\gamma = -1$ when $z_2 < 0$. In either case, we have

$$f'((x_1, 0); (z_1, z_2)) = \max_{\phi \in \partial f(x_1, 0)} \langle \phi, z \rangle.$$

The directional derivative function $f'(x; z)$ is linear for all z when x_2 is not zero, and when $x_2 = 0$, $f'(x; z)$ is linear for z in the subspace S of all z with $z_2 = 0$.

9.6 Functions and Operators

A function $F : \mathbb{R}^J \to \mathbb{R}^J$ is also called an *operator* on $\mathbb{R}^J$. For our purposes, the most important examples of operators on $\mathbb{R}^J$ are the orthogonal projections P_C onto convex sets, and gradient operators, that is, $F(x) = \nabla g(x)$, for some differentiable function $g(x) : \mathbb{R}^J \to \mathbb{R}$. As we shall see later, the operators P_C are also gradient operators.

Definition 9.16 *An operator $F(x)$ on $\mathbb{R}^J$ is called* L-Lipschitz continuous, *with respect to a given norm on $\mathbb{R}^J$, if, for every x and y in $\mathbb{R}^J$, we have*

$$\|F(x) - F(y)\| \leq L\|x - y\|.$$

Definition 9.17 *An operator $F(x)$ on $\mathbb{R}^J$ is called* nonexpansive, *with respect to a given norm on $\mathbb{R}^J$, if, for every x and y in $\mathbb{R}^J$, we have*

$$\|F(x) - F(y)\| \leq \|x - y\|.$$

Clearly, if an operator $F(x)$ is L-Lipschitz continuous, then the operator $G(x) = \frac{1}{L}F(x)$ is nonexpansive.

Definition 9.18 *An operator $F(x)$ on $\mathbb{R}^J$ is called* firmly nonexpansive, *with respect to the Euclidean norm on $\mathbb{R}^J$, if, for every x and y in $\mathbb{R}^J$, we have*

$$\langle F(x) - F(y), x - y\rangle \geq \|F(x) - F(y)\|_2^2.$$

Lemma 9.2 *A firmly nonexpansive operator on $\mathbb{R}^J$ is nonexpansive.*

We have the following analog of Theorem 9.8.

Theorem 9.20 *Let $h(x)$ be convex and differentiable and its derivative, $\nabla h(x)$, nonexpansive in the two-norm, that is,*

$$||\nabla h(b) - \nabla h(a)||_2 \leq ||b - a||_2,$$

for all a and b. Then $\nabla h(x)$ is firmly nonexpansive, which means that

$$\langle \nabla h(b) - \nabla h(a), b - a\rangle \geq ||\nabla h(b) - \nabla h(a)||_2^2.$$

Suppose that $g(x) : \mathbb{R}^J \to \mathbb{R}$ is convex and the function $F(x) = \nabla g(x)$ is L-Lipschitz. Let $h(x) = \frac{1}{L}g(x)$, so that $\nabla h(x)$ is a nonexpansive operator. According to Theorem 9.20, the operator $\nabla h(x) = \frac{1}{L}\nabla g(x)$ is firmly nonexpansive.

Unlike the proof of Theorem 9.8, the proof of Theorem 9.20 is not trivial. In [119] Golshtein and Tretyakov prove the following theorem, from which Theorem 9.20 follows immediately. The proof given here of Theorem 9.21 is different from that given in [119].

Theorem 9.21 *Let $g : \mathbb{R}^J \rightarrow \mathbb{R}$ be convex and differentiable. The following are equivalent:*

$$||\nabla g(x) - \nabla g(y)||_2 \leq ||x - y||_2; \tag{9.9}$$

$$g(x) \geq g(y) + \langle \nabla g(y), x - y \rangle + \frac{1}{2}||\nabla g(x) - \nabla g(y)||_2^2; \tag{9.10}$$

and

$$\langle \nabla g(x) - \nabla g(y), x - y \rangle \geq ||\nabla g(x) - \nabla g(y)||_2^2.$$

Proof: The only non-trivial step in the proof is showing that Inequality (9.9) implies Inequality (9.10). From Theorem 9.16 we see that Inequality (9.9) implies that the function $h(x) = \frac{1}{2}\|x\|_2^2 - g(x)$ is convex, and that

$$\frac{1}{2}\|x - y\|_2^2 \geq g(x) - g(y) - \langle \nabla g(y), x - y \rangle,$$

for all x and y. Now fix y and define

$$d(z) = D_g(z, y) = g(z) - g(y) - \langle \nabla g(y), z - y \rangle,$$

for all z. Since the function $g(z)$ is convex, so is $d(z)$. Since

$$\nabla d(z) = \nabla g(z) - \nabla g(y),$$

it follows from Inequality (9.9) that

$$\|\nabla d(z) - \nabla d(x)\|_2 \leq \|z - x\|_2,$$

for all x and z. Then, from our previous calculations, we may conclude that

$$\frac{1}{2}\|z - x\|_2^2 \geq d(z) - d(x) - \langle \nabla d(x), z - x \rangle,$$

for all z and x.

Now let x be arbitrary and

$$z = x - \nabla g(x) + \nabla g(y).$$

Then

$$0 \leq d(z) \leq d(x) - \frac{1}{2}\|\nabla g(x) - \nabla g(y)\|_2^2.$$

This completes the proof. ∎

We know from Corollary 9.1 that the function

$$g(x) = \frac{1}{2}\Big(\|x\|_2^2 - \|x - P_C x\|_2^2\Big)$$

is convex. As Corollary 12.1 tells us, its gradient is $\nabla g(x) = P_C x$. We showed in Corollary 4.1 that the operator P_C is nonexpansive by showing that it is actually firmly nonexpansive. Therefore, Theorem 9.20 can be viewed as a generalization of Corollary 4.1.

If $g(x)$ is convex and $f(x) = \nabla g(x)$ is L-Lipschitz, then $\frac{1}{L}\nabla g(x)$ is nonexpansive, so, by Theorem 9.20, it is firmly nonexpansive. It follows that, for $\gamma > 0$, the operator

$$Tx = x - \gamma \nabla g(x)$$

is averaged, whenever $0 < \gamma < \frac{2}{L}$. By Theorem 14.2, the iterative sequence $x^{k+1} = Tx^k = x^k - \gamma \nabla g(x^k)$ converges to a minimizer of $g(x)$, whenever minimizers exist.

9.7 Convex Sets and Convex Functions

In Chapter 4 we said that a function $f : \mathbb{R}^J \to (-\infty, \infty]$ is convex if and only if its epigraph is a convex set in $\mathbb{R}^{J+1}$. At the same time, every closed convex set $C \subseteq \mathbb{R}^J$ has the form

$$C = \{x | f(x) \leq 0\}, \tag{9.11}$$

for some convex function $f : \mathbb{R}^J \to \mathbb{R}$. We are tempted to assume that the smoothness of the function f will be reflected in the geometry of the set C. In particular, we may well expect that, if x is on the boundary of C and f is differentiable at x, then there is a unique hyperplane supporting C at x and its normal is $\nabla f(x)$; but this is wrong. Any closed convex nonempty set C can be written as in Equation (9.11), for the differentiable function

$$f(x) = \frac{1}{2}\|x - P_C x\|^2.$$

As we shall see later, the gradient of $f(x)$ is $\nabla f(x) = x - P_C x$, so that $\nabla f(x) = 0$ for every x in C. Nevertheless, the set C may have a unique supporting hyperplane at each boundary point, or it may have multiple such hyperplanes, regardless of the properties of the f used to define C.

When we first encounter gradients, usually in Calculus III, they are almost always described geometrically as a vector that is a normal for the hyperplane that is tangent to the level surface of f at that point, and as indicating the direction of greatest increase of f. However, this is not always the case.

Consider the function $f : \mathbb{R}^2 \to \mathbb{R}$ given by

$$f(x_1, x_2) = \frac{1}{2}\left(\sqrt{x_1^2 + x_2^2} - 1\right)^2,$$

for $x_1^2 + x_2^2 \geq 1$, and zero, otherwise. This function is differentiable and

$$\nabla f(x) = \frac{\|x\|_2 - 1}{\|x\|_2} x,$$

for $\|x\|_2 \geq 1$, and $\nabla f(x) = 0$, otherwise. The level surface in $\mathbb{R}^2$ of all x such that $f(x) \leq 0$ is the closed unit ball; it is not a simple closed curve. At every point of its boundary the gradient is zero, and yet at each boundary point there is a unique supporting tangent line.

Consider the function $f : \mathbb{R}^2 \to \mathbb{R}$ given by $f(x) = f(x_1, x_2) = x_1^2$. The level curve $C = \{x | f(x) = 0\}$ is the x_2 axis. For any x such that $x_1 = 0$ the hyperplane supporting C at x is C itself, and any vector of the form $(\gamma, 0)$ is a normal to C. But the gradient of $f(x)$ is zero at all points of C. So the gradient of f is not a normal vector to the supporting hyperplane.

9.8 Exercises

Ex. 9.1 *Say that a function $f : \mathbb{R} \to \mathbb{R}$ has the* intermediate value property *(IVP) if, for every a and b in $\mathbb{R}$ and, for any d between $f(a)$ and $f(b)$, there is c between a and b with $f(c) = d$. Let $g : \mathbb{R} \to \mathbb{R}$ be differentiable and $f(x) = g'(x)$. Show that f has the IVP, even if f is not continuous.*

Ex. 9.2 *Prove Proposition 9.1.*

Ex. 9.3 *Prove Proposition 9.2. Hint: Fix $z \in \mathbb{R}^J$ and show that the function $g(x) = f(x) - \langle z, x \rangle$ has a global minimizer.*

Ex. 9.4 *Let $g : \mathbb{R} \to \mathbb{R}$ be differentiable at $x = x_0$. Show that, if the line $y = mx + b$ passes through the point $(x_0, g(x_0))$ and $mx + b \leq g(x)$ for all x, then $m = g'(x_0)$.*

Ex. 9.5 *Prove Proposition 9.3.*

Ex. 9.6 *Prove Proposition 9.4.*

Ex. 9.7 *Prove Proposition 9.7.*

Ex. 9.8 *Prove Lemma 9.1.*

Ex. 9.9 *Let C be a nonempty convex subset of $\mathbb{R}^J$. Show that the core of C and the interior of C are the same. Hints: We need only consider the case in which the core of C is not empty. By shifting C if necessary, we* *may assume that 0 is in the core of C. Then we want to show that 0 is in the interior of C. The* gauge function *for C is*

$$\gamma_C(x) = \inf\{\lambda \geq 0 \,|x \in \lambda C\}.$$

Show that the interior of C is the set of all x for which $\gamma_C(x) < 1$.

Ex. 9.10 *Let* $p : \mathbb{R}^J \to \mathbb{R}$ *be sub-linear, and* $p(-x_n) = -p(x_n)$ *for* $n = 1, 2, ..., N$. *Show that p is linear on the span of* $\{x_1, ..., x_N\}$.

Ex. 9.11 *Prove Lemma 9.2.*

Ex. 9.12 *Show that, if* $\hat{x}$ *minimizes the function* $g(x)$ *over all x in* $\mathbb{R}^J$, *then* $u = 0$ *is in the sub-differential* $\partial g(\hat{x})$.

Ex. 9.13 *If* $f(x)$ *and* $g(x)$ *are convex functions on* $\mathbb{R}^J$, *is* $f(x) + g(x)$ *convex? Is* $f(x)g(x)$ *convex?*

Ex. 9.14 *Let* $\iota_C(x)$ *be the indicator function of the closed convex set C. Show that the sub-differential of the function* ι_C *at a point c in C is the normal cone to C at the point c, that is,* $\partial \iota_C(c) = N_C(c)$, *for all c in C.*

Ex. 9.15 *[201] Let* $g(t)$ *be a strictly convex function for* $t > 0$. *For* $x > 0$ *and* $y > 0$, *define the function*

$$f(x, y) = xg\Big(\frac{y}{x}\Big).$$

Use induction to prove that

$$\sum_{n=1}^{J} f(x_n, y_n) \geq f(x_+, y_+),$$

for any positive numbers x_n *and* y_n, *where* $x_+ = \sum_{n=1}^{J} x_n$. *Also show that equality obtains if and only if the finite sequences* $\{x_n\}$ *and* $\{y_n\}$ *are proportional.*

Ex. 9.16 *Use the result in Exercise 9.15 to obtain Cauchy's Inequality. Hint: Let* $g(t) = -\sqrt{t}$.

Ex. 9.17 *Use the result in Exercise 9.15 to obtain Hölder's Inequality. Hint: Let* $g(t) = -t^{1/q}$.

Ex. 9.18 *Use the result in Exercise 9.15 to obtain Minkowski's Inequality. Hint: Let* $g(t) = -(t^{1/p} + 1)^p$.

Ex. 9.19 *Use the result in Exercise 9.15 to obtain* **Milne's Inequality***:*

$$x_+y_+ \geq \left(\sum_{n=1}^{J}(x_n + y_n)\right)\left(\sum_{n=1}^{J}\frac{x_n y_n}{x_n + y_n}\right).$$

Hint: Let $g(t) = -\frac{t}{1+t}$.

Ex. 9.20 *For real numbers* $x > 0$ *and* $y > 0$*, let* $f(x, y)$ *be the* **Kullback–Leibler function,**

$$f(x, y) = KL(x, y) = x\left(\log\frac{x}{y}\right) + y - x.$$

Use Exercise 9.15 to show that

$$\sum_{n=1}^{J} KL(x_n, y_n) \geq KL(x_+, y_+).$$

Ex. 9.21 *Let* $x > 0$ *and* $y > 0$ *be vectors in* $\mathbb{R}^J$ *with entries* x_n *and* y_n*, respectively, and let*

$$KL(x, y) = \sum_{n=1}^{J} KL(x_n, y_n)$$

be the **Kullback–Leibler distance** *from* y *to* x*. Let* $y_+ = \sum_{n=1}^{J} y_n > 0$*. Show that*

$$KL(x, y) = KL(x_+, y_+) + KL\left(x, \frac{x_+}{y_+}y\right).$$

Chapter 10

Convex Programming

10.1 Chapter Summary

Convex programming (CP) refers to the minimization of a convex function of one or several variables over a convex set. The convex set is often defined in terms of inequalities involving other convex functions. We begin by describing the basic problems of CP. We then discuss characterizations of the solutions given by the Karush–Kuhn–Tucker (KKT) Theorem, the concept of duality, and use these tools to solve certain CP problems.

10.2 The Primal Problem

Let f and g_i, $i = 1, ..., I$, be convex functions defined on a nonempty closed convex subset C of $\mathbb{R}^J$. The *primal problem* in *convex programming* (CP) is the following:

$$\text{minimize } f(x), \text{ subject to } g_i(x) \leq 0, \text{ for } i = 1, ..., I. \quad \text{(P)} \tag{10.1}$$

For notational convenience, we define $g(x) = (g_1(x), ..., g_I(x))$. Then P becomes

$$\text{minimize } f(x), \text{ subject to } g(x) \leq 0. \quad \text{(P)}$$

The *feasible set* for P is

$$F = \{x | g(x) \leq 0\},$$

and the members of F are called *feasible points* for P.

Definition 10.1 *The problem P is said to be* consistent *if F is not empty, and* super-consistent *if there is x in F with $g_i(x) < 0$ for all $i = 1, ..., I$. Such a point x is then called a* Slater point.

10.2.1 The Perturbed Problem

For each z in $\mathbb{R}^I$ let

$$MP(z) = \inf\{f(x) | x \in C, g(x) \leq z\},$$

and $MP = MP(0)$. The convex programming problem P(z) is to minimize the function $f(x)$ over x in C with $g(x) \leq z$. The feasible set for P(z) is

$$F(z) = \{x | g(x) \leq z\}.$$

We shall be interested in properties of the function $MP(z)$, in particular, how the function $MP(z)$ behaves as z moves away from $z = 0$.

For example, let $f(x) = x^2$; the minimum occurs at $x = 0$. Now consider the perturbed problem, minimize $f(x) = x^2$, subject to $x \leq z$. For $z \leq 0$, the minimum of the perturbed problem occurs at $x = z$, and we have $MP(z) = z^2$. For $z > 0$ the minimum of the perturbed problem is the global minimum, which is at $x = 0$, so $MP(z) = 0$. The global minimum of $MP(z)$ also occurs at $z = 0$.

We have the following theorem concerning the function $MP(z)$; see the exercises for related results.

Theorem 10.1 *The function $MP(z)$ is convex and its domain, the set of all z for which $F(z)$ is not empty, is convex. If P is super-consistent, then $z = 0$ is an interior point of the domain of $MP(z)$.*

Proof: See [176], Theorem 5.2.6. ■

From Theorem 9.18 we know that, if P is super-consistent, then there is a vector u such that

$$MP(z) \geq MP(0) + \langle u, z - 0 \rangle. \tag{10.2}$$

In fact, we can show that, in this case, $u \leq 0$. Suppose that $u_i > 0$ for some i. Since $z = 0$ is in the interior of the domain of $MP(z)$, there is $r > 0$ such that $F(z)$ is not empty for all z with $||z|| < r$. Let $w_j = 0$ for $j \neq i$ and $w_i = r/2$. Then $F(w)$ is not empty and $MP(0) \geq MP(w)$, since $F \subseteq F(w)$. But from Equation (10.2) we have

$$MP(w) \geq MP(0) + \frac{r}{2} u_i > MP(0).$$

This is a contradiction, and we conclude that $u \leq 0$.

10.2.2 The Sensitivity Vector and the Lagrangian

From now on we shall use $\lambda^* = -u$ instead of u. For any z we have

$$\langle \lambda^*, z \rangle \geq MP(0) - MP(z);$$

so for $z \geq 0$ we have $MP(z) \leq MP(0)$, and

$$\langle \lambda^*, z \rangle \geq MP(0) - MP(z) \geq 0. \tag{10.3}$$

The quantity $\langle \lambda^*, z \rangle$ measures how much $MP(z)$ changes as we increase z away from $z = 0$; for that reason, λ^* is called the *sensitivity vector*, as well as the vector of *Lagrange multipliers*.

The Lagrangian function for the problem P is the function

$$L(x, \lambda) = f(x) + \sum_{i=1}^{I} \lambda_i g_i(x) = f(x) + \langle \lambda, g(x) \rangle,$$

defined for all x in C and $\lambda \geq 0$.

For each fixed x in C, let

$$F(x) = \sup_{\lambda \geq 0} L(x, \lambda).$$

If x is feasible for P, then $f(x) \geq L(x, \lambda)$, for all $\lambda \geq 0$, so that $f(x) \geq F(x)$. On the other hand, since $f(x) = L(x, 0) \leq F(x)$, we can conclude that $f(x) = F(x)$ for all feasible x in C. If x is not feasible, then $F(x) = +\infty$. Consequently, minimizing $f(x)$ over all feasible x in C is equivalent to minimizing $F(x)$ over all x in C; that is, we have removed the constraint that x be feasible for P. In the next section we pursue this idea further.

10.3 From Constrained to Unconstrained

In addition to being a measure of the sensitivity of $MP(z)$ to changes in z, the vector λ^* can be used to convert the original constrained minimization problem P into an unconstrained one.

Theorem 10.2 *If the problem P has a sensitivity vector $\lambda^* \geq 0$, in particular, when P is super-consistent, then $MP(0) = \inf_{x \in C} L(x, \lambda^*)$, that is,*

$$MP(0) = \inf_{x \in C} \Big(f(x) + \langle \lambda^*, g(x) \rangle \Big). \tag{10.4}$$

Proof: For any fixed x in the set C, the set

$$F(g(x)) = \{t | g(t) \leq g(x)\}$$

contains $t = x$ and therefore is nonempty. By Equation (10.3)

$$MP(g(x)) + \langle \lambda^*, g(x) \rangle \geq MP(0).$$

Since x is in $F(g(x))$, we have

$$f(x) \geq MP(g(x)),$$

and it follows that

$$f(x) + \langle \lambda^*, g(x) \rangle \geq MP(0).$$

Therefore,

$$\inf_{x \in C} \Big(f(x) + \langle \lambda^*, g(x) \rangle \Big) \geq MP(0).$$

But

$$\inf_{x \in C} \Big(f(x) + \langle \lambda^*, g(x) \rangle \Big) \leq \inf_{x \in C, g(x) \leq 0} \Big(f(x) + \langle \lambda^*, g(x) \rangle \Big),$$

and

$$\inf_{x \in C, g(x) \leq 0} \Big(f(x) + \langle \lambda^*, g(x) \rangle \Big) \leq \inf_{x \in C, g(x) \leq 0} f(x) = MP(0),$$

since $\lambda^* \geq 0$ and $g(x) \leq 0$. ∎

Note that the theorem tells us that the two sides of Equation (10.4) are equal. Although it is true, we cannot conclude, from Theorem 10.2 alone, that if both sides have a minimizer then the minimizers are the same vector.

10.4 Saddle Points

To prepare for our discussion of the Karush–Kuhn–Tucker Theorem and duality, we consider the notion of *saddle points.*

10.4.1 The Primal and Dual Problems

Suppose that X and Y are two nonempty sets and $K : X \times Y \to (-\infty, \infty)$ is a function of two variables. For each x in X, define the function $f(x)$ by the *supremum*

$$f(x) = \sup_y K(x, y), \tag{10.5}$$

where the supremum, abbreviated "sup," is the least upper bound of the real numbers $K(x, y)$, over all y in Y. Then we have

$$K(x, y) \leq f(x), \tag{10.6}$$

for all x. Similarly, for each y in Y, define the function $g(y)$ by

$$g(y) = \inf_x K(x, y); \tag{10.7}$$

here the infimum is the greatest lower bound of the numbers $K(x, y)$, over all x in X. Then we have

$$g(y) \leq K(x, y), \tag{10.8}$$

for all y in Y. Putting together (10.6) and (10.8), we have

$$g(y) \le K(x,y) \le f(x),$$

for all x and y. Now we consider two problems: the *primal problem* is minimizing $f(x)$ and the *dual problem* is maximizing $g(y)$.

Definition 10.2 *The pair* $(\hat{x}, \hat{y})$ *is called a* saddle point *for the function* $K(x,y)$ *if, for all* x *and* y*, we have*

$$K(\hat{x}, y) \le K(\hat{x}, \hat{y}) \le K(x, \hat{y}).$$

The number $K(\hat{x}, \hat{y})$ *is called the* saddle value.

For example, the function $K(x,y) = x^2 - y^2$ has $(0,0)$ for a saddle point, with saddle value zero.

10.4.2 The Main Theorem

We have the following theorem, with the proof left to the reader.

Theorem 10.3 *The following are equivalent:*

(1) The pair $(\hat{x}, \hat{y})$ *forms a saddle point for* $K(x,y)$*;*

(2) The point $\hat{x}$ *solves the primal problem, that is,* $\hat{x}$ *minimizes* $f(x)$*, over all* x *in* X*, and* $\hat{y}$ *solves the dual problem, that is,* $\hat{y}$ *maximizes* $g(y)$*, over all* y *in* Y*, and* $f(\hat{x}) = g(\hat{y})$*.*

When $(\hat{x}, \hat{y})$ forms a saddle point for $K(x,y)$, we have

$$g(y) \le K(\hat{x}, \hat{y}) \le f(x),$$

for all x and y, so that the maximum value of $g(y)$ and the minimum value of $f(x)$ are both equal to $K(\hat{x}, \hat{y})$.

10.4.3 A Duality Approach to Optimization

Suppose that our original problem is to minimize a function $f(x)$ over x in some set X. One approach is to find a second set Y and a function $K(x,y)$ of two variables for which Equation (10.5) holds, use Equation (10.7) to construct a second function $g(y)$, defined for y in Y, and then maximize $g(y)$. If a saddle point exists, then, according to the theorem, we have solved the original problem.

10.5 The Karush–Kuhn–Tucker Theorem

We begin with sufficient conditions for a vector x^* to be a solution to the primal CP problem. Under certain restrictions, as specified by the Karush–Kuhn–Tucker Theorem, these conditions become necessary, as well.

10.5.1 Sufficient Conditions

Proposition 10.1 *Let x^* be a member of C. If there is $\lambda^* \geq 0$ such that, for all $x \in C$ and all vectors $\lambda \geq 0$, we have*

$$L(x^*, \lambda) \leq L(x^*, \lambda^*) \leq L(x, \lambda^*),$$

then x^* is feasible and x^* solves the primal CP problem.

Proof: The proof is left as Exercise 10.1. ∎

Corollary 10.1 *If, for a given vector $x^* \in C$, there is $\lambda^* \geq 0$ such that*

$$L(x^*, \lambda^*) \leq L(x, \lambda^*),$$

for all $x \in C$, and $\lambda_i^* g_i(x^*) = 0$, for all i, then x^* is feasible and x^* solves the primal CP problem.

Proof: The proof is left as Exercise 10.2. ∎

10.5.2 The KKT Theorem: Saddle-Point Form

This form of the KKT Theorem does not require that the functions involved be differentiable. The *saddle-point* form of the Karush–Kuhn–Tucker (KKT) Theorem is the following.

Theorem 10.4 *Let P, the primal CP problem, be super-consistent. Then x^* solves P if and only if there is a vector λ^* such that*

(1) $\lambda^* \geq 0$;

(2) $L(x^*, \lambda) \leq L(x^*, \lambda^*) \leq L(x, \lambda^*)$, *for all $x \in C$ and all $\lambda \geq 0$;*

(3) $\lambda_i^* g_i(x^*) = 0$, *for all $i = 1, ..., I$.*

Proof: Since P is super-consistent and x^* solves P, we know from Theorem 10.2 that there is $\lambda^* \geq 0$ such that

$$f(x^*) = \inf_{x \in C} L(x, \lambda^*).$$

We do not yet know that $f(x^*) = L(x^*, \lambda^*)$, however. We do have

$$f(x^*) \leq L(x^*, \lambda^*) = f(x^*) + \langle \lambda^*, g(x^*) \rangle,$$

though, and since $\lambda^* \geq 0$ and $g(x^*) \leq 0$, we also have

$$f(x^*) + \langle \lambda^*, g(x^*) \rangle \leq f(x^*).$$

Now we can conclude that $f(x^*) = L(x^*, \lambda^*)$ and $\langle \lambda^*, g(x^*) \rangle = 0$. It follows that $\lambda_i^* g_i(x^*) = 0$, for all $i = 1, ..., I$. Since, for $\lambda \geq 0$,

$$L(x^*, \lambda^*) - L(x^*, \lambda) = \langle \lambda^* - \lambda, g(x^*) \rangle = \langle -\lambda, g(x^*) \rangle \geq 0,$$

we also have

$$L(x^*, \lambda) \leq L(x^*, \lambda^*),$$

for all $\lambda \geq 0$.

Conversely, suppose that x^* and λ^* satisfy the three conditions of the theorem. First, we show that x^* is feasible for P, that is, $g(x^*) \leq 0$. Let i be fixed and take λ to have the same entries as λ^*, except that $\lambda_i = \lambda_i^* + 1$. Then we have $\lambda \geq 0$ and

$$0 \leq L(x^*, \lambda^*) - L(x^*, \lambda) = -g_i(x^*).$$

Also,

$$f(x^*) = L(x^*, 0) \leq L(x^*, \lambda^*) = f(x^*) + \langle \lambda^*, g(x^*) \rangle = f(x^*),$$

so

$$f(x^*) = L(x^*, \lambda^*) \leq L(x, \lambda^*).$$

But we also have

$$L(x^*, \lambda^*) \leq \inf_{x \in C} \Big(f(x) + \langle \lambda^*, g(x) \rangle \Big) \leq \inf_{x \in C, g(x) \leq 0} f(x).$$

We conclude that $f(x^*) = MP(0)$, and since x^* is feasible for P, x^* solves P. ∎

Condition (3) is called *complementary slackness*. If $g_i(x^*) = 0$, we say that the ith constraint is *binding*.

10.5.3 The KKT Theorem: The Gradient Form

Now we assume that the functions $f(x)$ and $g_i(x)$ are differentiable.

Theorem 10.5 *Let P be super-consistent. Then x^* solves P if and only if there is a vector λ^* such that*

(1) $\lambda^* \geq 0$;

(2) $\lambda_i^* g_i(x^*) = 0$, *for all* $i = 1, ..., I$;

(3) $\nabla f(x^*) + \sum_{i=1}^{I} \lambda_i^* \nabla g_i(x^*) = 0$.

The proof is similar to the previous one and we omit it. The interested reader should consult [176, p. 185].

10.6 On Existence of Lagrange Multipliers

As we saw previously, if P is super-consistent, then $z = 0$ is in the interior of the domain of the function $MP(z)$, and so the sub-differential of $MP(z)$ is nonempty at $z = 0$. The sub-gradient d was shown to be nonpositive and we defined the sensitivity vector, or the vector of Lagrange multipliers, to be $\lambda^* = -d$. Theorem 10.5 tells us that if P is super-consistent and x^* solves P, then the vector $\nabla f(x^*)$ is a nonnegative linear combination of the vectors $-\nabla g_i(x^*)$. This sounds like the assertion in Farkas' Lemma.

For any point x, define the set

$$B(x) = \{i | g_i(x) = 0\},$$

and

$$Z(x) = \{z | z^T \nabla g_i(x) \leq 0,\ i \in B(x), \text{ and } z^T \nabla f(x) < 0\}.$$

If $Z(x)$ is empty, then

$$z^T(-\nabla g_i(x)) \geq 0$$

for $i \in B(x)$ implies

$$z^T \nabla f(x) \geq 0,$$

which, by Farkas' Lemma, implies that $\nabla f(x)$ is a nonnegative linear combination of the vectors $-\nabla g_i(x)$ for $i \in B(x)$. The objective, then, is to find some condition which, if it holds at the solution x^*, will imply that $Z(x^*)$ is empty; first-order necessary conditions are of this sort. It will then follow that there are nonnegative Lagrange multipliers for which

$$\nabla f(x^*) + \sum_{i=1}^{I} \lambda_i^* \nabla g_i(x^*) = 0;$$

for i not in $B(x^*)$ we let $\lambda_i^* = 0$. For more discussion of this issue, see Fiacco and McCormick [112].

10.7 The Problem of Equality Constraints

We consider now what happens when some of the constraints are equalities.

10.7.1 The Problem

Let f and g_i, $i = 1, ..., I$, be differentiable functions defined on $\mathbb{R}^J$. We consider the following problem: Minimize $f(x)$, subject to the constraints $g_i(x) \leq 0$, for $i = 1, ..., K$, and $g_i(x) = 0$, for $i = K + 1, ..., I$. If $1 \leq K < I$, the constraints are said to be mixed. If $K = I$, there are only inequality constraints, so, for convex $f(x)$ and $g_i(x)$, the problem is P, given by (10.1). If $K < I$, we cannot convert it to a CP problem by rewriting the equality constraints as $g_i(x) \leq 0$ and $-g_i(x) \leq 0$, since then we would lose the convexity property of the constraint functions. Nevertheless, a version of the KKT Theorem holds for such problems.

Definition 10.3 *The* feasible set *for this problem is the set F of all x satisfying the constraints.*

Definition 10.4 *The problem is said to be* consistent *if F is not empty.*

Definition 10.5 *Let $\mathcal{I}(x)$ be the set of all indices $1 \leq i \leq I$ for which $g_i(x) = 0$. The point x is* regular *if the set of gradients $\{\nabla g_i(x) | i \in \mathcal{I}(x)\}$ is linearly independent.*

10.7.2 The KKT Theorem for Mixed Constraints

The following version of the KKT Theorem provides a necessary condition for a regular point x^* to be a local constrained minimizer.

Theorem 10.6 *Let x^* be a regular point for the problem described in Subsection 10.7.1. If x^* is a local constrained minimizer of $f(x)$, then there is a vector λ^* such that*

(1) $\lambda_i^* \geq 0$, for $i = 1, ..., K$;

(2) $\lambda_i^* g_i(x^*) = 0$, for $i = 1, ..., K$;

(3) $\nabla f(x^*) + \sum_{i=1}^{I} \lambda_i^* \nabla g_i(x^*) = 0$.

Note that, if there are some equality constraints, then the vector λ need not be nonnegative.

10.7.3 The KKT Theorem for LP

Consider the LP problem PS: Minimize $z = c^T x$, subject to $Ax = b$ and $x \geq 0$. We let

$$z = f(x) = c^T x,$$

$$g_i(x) = b_i - (Ax)_i,$$

for $i = 1, ..., I$, and

$$g_i(x) = -x_j,$$

for $i = I + 1, ..., I + J$ and $j = i - I$. We assume that $I < J$ and that the I by J matrix A has rank I. Then, since $-\nabla g_i(x)$ is a^i, the ith column of A^T, the vectors $\{\nabla g_i(x) \,|\, i = 1, ..., I\}$ are linearly independent and every $x > 0$ is a regular point.

Suppose that a regular point x^* solves PS. Let λ^* be the vector in $\mathbb{R}^{I+J}$ whose existence is guaranteed by Theorem 10.6. Denote by y^* the vector in $\mathbb{R}^I$ whose entries are the first I entries of λ^*, and r the nonnegative vector in $\mathbb{R}^J$ whose entries are the last J entries of λ^*. Then, applying Theorem 10.6, we have $r^T x^* = 0$, $Ax^* = b$, and

$$c - \sum_{i=1}^{I} \lambda_i^* a^i + \sum_{j=1}^{J} r_j(-\delta^j) = 0,$$

or,

$$c - A^T y^* = r \geq 0,$$

where δ^j is the column vector whose jth entry is one and the rest are zero.

The KKT Theorem for this problem is then the following.

Theorem 10.7 *Let A have full rank I. The regular point x^* solves PS if and only if there are vectors y^* in $\mathbb{R}^I$ and $r \geq 0$ in $\mathbb{R}^J$ such that*

(1) $Ax^* = b$;

(2) $r = c - A^T y^*$;

(3) $r^T x^* = 0$.

Then y^* solves DS.

The first condition in the theorem is *primal feasibility*, the second one is *dual feasibility*, and the third is *complementary slackness*. The first two conditions tell us that x^* is feasible for PS and y^* is feasible for DS. Combining these two conditions with complementary slackness, we can write

$$z^* = c^T x^* = (A^T y^* + r)^T x^* = (A^T y^*)^T x^* + r^T x^* = (y^*)^T b = w^*,$$

so $z^* = w^*$ and there is no duality gap. Invoking Corollary 6.2 to the Weak Duality Theorem, we conclude that x^* and y^* solve their respective problems.

10.7.4 The Lagrangian Fallacy

As Kalman notes in [135], it is quite common, when discussing the use of Lagrange multipliers in optimization, to say, incorrectly, that the problem of minimizing $f(x)$, subject to $g(x) = 0$, has been converted into the problem of finding a local minimum of the Lagrangian function $L(x, \lambda)$, as a function of (x, λ). The following example, taken from [135], shows that this interpretation is false.

Minimize the function $f(x, y) = x^2 + y^2$, subject to $g(x, y) = xy - 1 = 0$. Using a Lagrange multiplier λ, and the Lagrangian

$$L(x, y, \lambda) = x^2 + y^2 + \lambda(xy - 1) = (x - y)^2 + \lambda(xy - 1) + 2xy,$$

we find that

$$2x + \lambda y = 0,$$

$$2y + \lambda x = 0,$$

and

$$xy - 1 = 0.$$

It follows that $x = 1$, $y = 1$, $\lambda = -2$, and $L(1, 1, -2) = 2$. Now let us move away from the point $(1, 1, -2)$ along the line $(x, x, -2 + t)$, so that the Lagrangian takes on the values

$$L(x, x, -2 + t) = (x - x)^2 + (-2 + t)(x^2 - 1) + 2x^2 = 2 + t(x^2 - 1).$$

For small positive values of t, the Lagrangian takes on values greater than 2, while, for small negative values of t, its values are smaller than 2.

10.8 Two Examples

We illustrate the use of the gradient form of the KKT Theorem with two examples that appeared in the paper of Driscoll and Fox [100].

10.8.1 A Linear Programming Problem

Minimize $f(x_1, x_2) = 3x_1 + 2x_2$, subject to the constraints $2x_1 + x_2 \geq 100$, $x_1 + x_2 \geq 80$, $x_1 \geq 0$ and $x_2 \geq 0$. We define

$$\begin{aligned} g_1(x_1, x_2) &= 100 - 2x_1 - x_2, \\ g_2(x_1, x_2) &= 80 - x_1 - x_2, \\ g_3(x_1, x_2) &= -x_1, \text{ and} \\ g_4(x_1, x_2) &= -x_2. \end{aligned}$$

The Lagrangian is then

$$L(x,\lambda) = 3x_1 + 2x_2 + \lambda_1(100 - 2x_1 - x_2) + \lambda_2(80 - x_1 - x_2) - \lambda_3 x_1 - \lambda_4 x_2.$$

From the KKT Theorem, we know that, if there is a solution x^*, then there is $\lambda^* \geq 0$ with

$$f(x^*) = L(x^*, \lambda^*) \leq L(x, \lambda^*),$$

for all x. For notational simplicity, we write λ in place of λ^*.

Taking the partial derivatives of $L(x,\lambda)$ with respect to the variables x_1 and x_2, we get

$$3 - 2\lambda_1 - \lambda_2 - \lambda_3 = 0, \quad \text{and} \quad 2 - \lambda_1 - \lambda_2 - \lambda_4 = 0.$$

The complementary slackness conditions are

$$\begin{aligned} \lambda_1 &= 0, \quad \text{if} \quad 2x_1 + x_2 \neq 100, \\ \lambda_2 &= 0, \quad \text{if} \quad x_1 + x_2 \neq 80, \\ \lambda_3 &= 0, \quad \text{if} \quad x_1 \neq 0, \\ \lambda_4 &= 0, \quad \text{if} \quad x_2 \neq 0. \end{aligned}$$

A little thought reveals that precisely two of the four constraints must be binding. Examining the six cases, we find that the only case satisfying all the conditions of the KKT Theorem is $\lambda_3 = \lambda_4 = 0$. The minimum occurs at $x_1 = 20$ and $x_2 = 60$ and the minimum value is $f(20, 60) = 180$.

We can use these results to illustrate Theorem 10.2. The sensitivity vector is $\lambda^* = (1, 1, 0, 0)$ and the Lagrangian function at λ^* is

$$L(x, \lambda^*) = 3x_1 + 2x_2 + 1(100 - 2x_1 - x_2) + 1(80 - x_1 - x_2).$$

In this case, we find that $L(x, \lambda^*) = 180$, for all x.

10.8.2 A Nonlinear Convex Programming Problem

Minimize the function

$$f(x_1, x_2) = (x_1 - 14)^2 + (x_2 - 11)^2,$$

subject to

$$g_1(x_1, x_2) = (x_1 - 11)^2 + (x_2 - 13)^2 - 49 \leq 0,$$

and

$$g_2(x_1, x_2) = x_1 + x_2 - 19 \leq 0.$$

The Lagrangian is then

$$L(x,\lambda) = (x_1 - 14)^2 + (x_2 - 11)^2 + \lambda_1\Big((x_1 - 11)^2 + (x_2 - 13)^2 - 49\Big) + \lambda_2\Big(x_1 + x_2 - 19\Big).$$

Again, we write λ in place of λ^*. Setting the partial derivatives, with respect to x_1 and x_2, to zero, we get the KKT equations

$$2x_1 - 28 + 2\lambda_1 x_1 - 22\lambda_1 + \lambda_2 = 0,$$

and

$$2x_2 - 22 + 2\lambda_1 x_2 - 26\lambda_1 + \lambda_2 = 0.$$

The complementary slackness conditions are

$$\lambda_1 = 0, \quad \text{if} \quad (x_1 - 11)^2 + (x_2 - 13)^2 \neq 49,$$
$$\lambda_2 = 0, \quad \text{if} \quad x_1 + x_2 \neq 19.$$

There are four cases to consider. First, if neither constraint is binding, the KKT equations have solution $x_1 = 14$ and $x_2 = 11$, which is not feasible. If only the first constraint is binding, we obtain two solutions, neither feasible. If only the second constraint is binding, we obtain $x_1^* = 11$, $x_2^* = 8$, and $\lambda_2 = 6$. This is the optimal solution. If both constraints are binding, we obtain, with a bit of calculation, two solutions, neither feasible. The minimum value is $f(11, 8) = 18$, and the sensitivity vector is $\lambda^* = (0, 6)$. Using these results, we once again illustrate Theorem 10.2.

The Lagrangian function at λ^* is

$$L(x, \lambda^*) = (x_1 - 14)^2 + (x_2 - 11)^2 + 6(x_1 + x_2 - 19).$$

Setting to zero the first partial derivatives of $L(x, \lambda^*)$, we get

$$0 = 2(x_1 - 14) + 6,$$

and

$$0 = 2(x_2 - 11) + 6,$$

so that $x_1^* = 11$ and $x_2^* = 8$. Note that Theorem 10.2 only guarantees that 18 is the infimum of the function $L(x, \lambda^*)$. It does not say that this smallest value must occur at $x = x^*$ or even occurs anywhere; that is, it does not say that $L(x^*, \lambda^*) \leq L(x, \lambda^*)$. This stronger result comes from the KKT Theorem.

In this problem, we are able to use the KKT Theorem and a case-by-case analysis to find the solution because the problem is artificial, with few variables and constraints. In practice there will be many more variables and constraints, making such a case-by-case approach impractical. It is for that reason that we turn to iterative optimization methods.

10.9 The Dual Problem

The *dual problem* (DP) corresponding to P is to maximize

$$h(\lambda) = \inf_{x \in C} L(x, \lambda),$$

for $\lambda \geq 0$. Let

$$MD = \sup_{\lambda \geq 0} h(\lambda). \tag{10.9}$$

A vector $\lambda \geq 0$ is feasible for DP if $h(\lambda) > -\infty$. Then DP is consistent if there are feasible λ. Recall that Theorem 10.2 tells us that if a sensitivity vector $\lambda^* \geq 0$ exists, then $h(\lambda^*) = MP$.

10.9.1 When Is $MP = MD$?

We have the following theorem.

Theorem 10.8 *Assume that P is super-consistent, so that there is a sensitivity vector $\lambda^* \geq 0$, and that MP is finite. Then*

(1) $MP = MD$;

(2) $MD = h(\lambda^*)$, so the supremum in Equation (10.9) is attained at λ^*;

(3) if the infimum in the definition of MP is attained at x^*, then $\langle \lambda^*, g(x^*) \rangle = 0$;

(4) such an x^* also minimizes $L(x, \lambda^*)$ over $x \in C$.

Proof: For all $\lambda \geq 0$ we have

$$h(\lambda) = \inf_{x \in C} L(x, \lambda) \leq \inf_{x \in C, g(x) \leq 0} L(x, \lambda) \leq \inf_{x \in C, g(x) \leq 0} f(x) = MP.$$

Therefore, $MD \leq MP$. The difference $MP - MD$ is known as the *duality gap* for CP. We also know that

$$MP = h(\lambda^*) \leq MD,$$

so $MP = MD$, and the supremum in the definition of MD is attained at λ^*. From

$$f(x^*) = MP = \inf_{x \in C} L(x, \lambda^*) \leq \inf_{x \in C, g(x) \leq 0} L(x, \lambda^*)$$

$$\leq L(x^*, \lambda^*) \leq f(x^*),$$

it follows that $\langle \lambda^*, g(x^*) \rangle = 0$. ∎

10.9.2 The Primal-Dual Method

From Theorem 10.8 we see that one approach to solving P is to solve DP for λ^* and then minimize $L(x, \lambda^*)$ over $x \in C$. This is useful only if solving DP is simpler than solving P directly. Each evaluation of $h(\lambda)$ involves minimizing $L(x, \lambda)$ over $x \in C$. Once we have found λ^*, we find x^* by minimizing $L(x, \lambda^*)$ over $x \in C$. The advantage is that all the minimizations are over all $x \in C$, not over just the feasible vectors.

10.9.3 Using the KKT Theorem

As we noted previously, using the KKT Theorem and a case-by-case analysis, as in the example problems, is not practical for real-world problems involving many variables and constraints. The KKT Theorem can, however, tell us something about the nature of the solution, and perhaps help us design an algorithm to solve the problem, as the following two examples illustrate.

10.10 Nonnegative Least-Squares Solutions

If there is no solution to a system of linear equations $Ax = b$, then we may seek a *least-squares* "solution," which is a minimizer of the function

$$f(x) = \sum_{i=1}^{I} \left(\left(\sum_{m=1}^{J} A_{im} x_m \right) - b_i \right)^2 = ||Ax - b||_2^2.$$

The partial derivative of $f(x)$ with respect to the variable x_j is

$$\frac{\partial f}{\partial x_j}(x) = 2 \sum_{i=1}^{I} A_{ij} \left(\left(\sum_{m=1}^{J} A_{im} x_m \right) - b_i \right).$$

Setting the gradient equal to zero, we find that to get a least-squares solution we must solve the system of equations

$$A^T(Ax - b) = 0.$$

Now we consider what happens when the additional constraints $x_j \geq 0$ are imposed.

This problem fits into the CP framework, when we define

$$g_j(x) = -x_j,$$

for each j. Let $\hat{x}$ be a least-squares solution. According to the KKT Theorem, for those values of j for which $\hat{x}_j$ is not zero we have $\lambda_j^* = 0$ and $\frac{\partial f}{\partial x_j}(\hat{x}) = 0$. Therefore, if $\hat{x}_j \neq 0$,

$$0 = \sum_{i=1}^{I} A_{ij}\Bigg(\Big(\sum_{m=1}^{J} A_{im}\hat{x}_m\Big) - b_i\Bigg).$$

Let Q be the matrix obtained from A by deleting columns j for which $\hat{x}_j = 0$. Then we can write

$$Q^T(A\hat{x} - b) = 0.$$

If the matrix Q has full rank, which will almost always be the case, and has at least I columns, then Q^T is a one-to-one linear transformation, which implies that $A\hat{x} = b$. Therefore, when there is no nonnegative solution of $Ax = b$, Q must have fewer than I columns, which means that $\hat{x}$ has fewer than I nonzero entries. We can state this result more formally.

Definition 10.6 *The matrix A has the* full-rank property *if A and every matrix Q obtained from A by deleting columns have full rank.*

Theorem 10.9 *Let A have the full-rank property. Suppose there is no nonnegative solution to the system of equations $Ax = b$. Then there is a subset S of the set $\{j = 1, 2, ..., J\}$, with cardinality at most $I - 1$, such that, if $\hat{x}$ is any minimizer of $||Ax - b||_2$ subject to $x \geq 0$, then $\hat{x}_j = 0$ for j not in S. Therefore, $\hat{x}$ is unique.*

This result has some practical implications in medical image reconstruction.

10.11 An Example in Image Reconstruction

In many areas of image processing, including medical imaging, the vector x is a vectorized image that we seek, whose typically nonnegative entries are the unknown pixel values, the entries of b are measurements obtained through the use of some device, such as a CAT-scan, and the matrix A describes, usually imperfectly, the relationship between the desired image x and the data b. In transmission tomography the data is often viewed as integrals along line segments through the object; in the discrete version, the data may be viewed as the sums of the x_j for those j for which the associated pixel intersects the given line segment. A crude estimate of the matrix A is to define $A_{i,j} = 1$ if the ith line segment intersects the jth pixel, and $A_{i,j} = 0$ otherwise. Note that the matrix A is *sparse*, that is, most of its entries are zero. This is typical of such remote-sensing problems.

It is helpful to note that the matrix A as just presented does not do a very good job of describing how the data is related to the pixels. By using only the values zero or one, we ignore the obvious fact that a line segment may intersect most of one pixel, while touching only a little of another. We need to improve A, if we are to reduce the *model error*. We can do a better job by taking the entries of A to be numbers between zero and one that are the relative sizes of the intersection of the given line segment with the given pixel.

There are other sources of error, as well: the line-integral model is only an approximation; x-rays do not travel along exact straight lines, but along narrow strips; the frequency content of the rays can change as the rays travel through the body; the measured data are not precisely the sums given by the vector Ax, regardless of how accurately we describe the intersection of the line segments with the pixels. In short, the vector b also contains *noise*, known as *measurement noise*. For all these reasons, there may not be exact nonnegative solutions of $Ax = b$, and even if there are such solutions, they may not be suitable for diagnosis.

Once the data is obtained, the number of measurements I is determined. The number of pixels J is not yet fixed, and we can select J to suit our needs. The scene being imaged or the patient being scanned has no pixels; these are artificially imposed by us. If J is too small, we will not obtain the desired resolution in the reconstructed image.

In the hope of improving the resolution of the reconstructed image, we may be tempted to take J, the number of pixels, larger than I, the number of equations arising from our measurement. Since the vector b consists of measured data, it is noisy and there may well not be a nonnegative exact solution of $Ax = b$. As a result, the image obtained by nonnegatively constrained least-squares will have at most $I - 1$ nonzero entries; many of the pixels will be zero and they will be scattered throughout the image, making it unusable. The reconstructed images resemble stars in a night sky, and, as a result, the theorem is sometimes described as the "night sky" theorem.

This "night sky" phenomenon is not restricted to least squares. The same thing happens with methods based on the Kullback–Leibler distance, such as MART, EMML and SMART [39]. These algorithms are discussed in some detail in Chapter 11.

10.12 Solving the Dual Problem

In this section we use the KKT Theorem to derive an iterative algorithm to minimize the function

$$f(x) = \frac{1}{2}||x||_2^2,$$

subject to $Ax \geq b$, by solving the dual problem of maximizing $h(\lambda)$, over $\lambda \geq 0$.

10.12.1 The Primal and Dual Problems

Minimizing $f(x)$ over x such that $Ax \geq b$ is the primal problem. Here we let $g_i = b_i - (Ax)_i$, for $i = 1, ..., I$, and the set C be all of $\mathbb{R}^J$. The Lagrangian is then

$$L(x, \lambda) = \frac{1}{2}||x||_2^2 - \lambda^T Ax + \lambda^T b.$$

The infimum of $L(x, \lambda)$ over all x occurs when $x = A^T\lambda$ and so

$$h(\lambda) = \lambda^T b - \frac{1}{2}||A^T\lambda||_2^2.$$

For any x satisfying $Ax \geq b$ and any $\lambda \geq 0$ we have $h(\lambda) \leq f(x)$. If x^* is the unique solution of the primal problem and λ^* any solution of the dual problem, we have $f(x^*) = h(\lambda^*)$. The point here is that the constraints in the dual problem are easier to implement in an iterative algorithm, so solving the dual problem is the simpler task.

The algorithm we present now calculates iteratively two sequences, $\{x^k\}$ and $\{\lambda^k\}$, such that $f(x^k) - h(\lambda^k)$ converges to zero. The limits of $\{x^k\}$ and $\{\lambda^k\}$ will be the solutions of the primal and dual problems, respectively.

10.12.2 Hildreth's Dual Algorithm

The iterative algorithm we describe here was originally published by Hildreth [128], and later extended by Lent and Censor [148]. It is a *row-action* method in that, at each step of the iteration, only a single row of the matrix A is used. Having found x^k and λ^k, we use $i = k(\text{mod}\, I) + 1$, A_i the i-th row of A, and b_i to calculate x^{k+1} and λ^{k+1}.

We know that the optimal x^* and $\lambda^* \geq 0$ must satisfy $x^* = A^T\lambda^*$. Therefore, the algorithm guarantees that, at each step, we have $\lambda^k > 0$ and $x^k = A^T\lambda^k$.

Having found x^k and λ^k, we proceed as follows. First, we select $i = k(\text{mod}\, I) + 1$. Since

$$h(\lambda) = b^T\lambda - \frac{1}{2}||A^T\lambda||_2^2,$$

we have

$$\nabla h(\lambda) = b - AA^T\lambda.$$

A gradient ascent method to maximize $h(\lambda)$ would then have the iterative step

$$\lambda^{k+1} = \lambda^k + \gamma_k(b - AA^T\lambda^k) = \lambda^k + \gamma_k(b - Ax^k),$$

for some $\gamma_k > 0$. A row-action variant of gradient ascent modifies only the i-th entry of λ at the k-th step, with

$$\lambda_i^{k+1} = \lambda_i^k + \gamma_k(b_i - (Ax^k)_i).$$

Since we require that $\lambda^{k+1} \geq 0$, when $(b_i - (Ax^k)_i) < 0$ we must select γ_k so that

$$\gamma_k(b_i - (Ax^k)_i) \geq -\lambda_i^k.$$

We then have

$$x^{k+1} = x^k + \gamma_k(b_i - (Ax^k)_i)A_i^T,$$

which is used in the next step, in forming $\nabla h(\lambda^{k+1})$. Proof of convergence of this algorithm is presented in [83].

10.13 Minimum One-Norm Solutions

When the system of linear equations $Ax = b$ is under-determined, it is common practice to seek a solution that also minimizes some objective function. For example, the *minimum two-norm solution* is the vector x satisfying $Ax = b$ for which the (square of the) two-norm,

$$||x||_2^2 = \sum_{j=1}^{J} x_j^2,$$

is minimized. Alternatively, we may seek the *minimum one-norm solution*, for which the one-norm,

$$||x||_1 = \sum_{j=1}^{J} |x_j|,$$

is minimized.

If the vector x is required to be nonnegative, then the one-norm is simply the sum of the entries, and minimizing the one-norm subject to

$Ax = b$ becomes a linear programming problem. This is the situation in applications involving image reconstruction.

In *compressed sensing* and *compressed sampling* [98] one seeks a solution of $Ax = b$ having the minimal number of nonzero entries; the vector x here is not assumed to be nonnegative. Under certain restrictions on the matrix A, the solution can be found by minimizing the one-norm of x, subject to the constraints $Ax = b$. The one-norm is not a linear functional of x, but the problem can still be converted into a linear programming problem, as we shall see next.

10.13.1 Reformulation as an LP Problem

The entries of x need not be nonnegative, so the problem is not yet a linear programming problem. Let

$$B = \begin{bmatrix} A & -A \end{bmatrix},$$

and consider the linear programming problem of minimizing the function

$$c^T z = \sum_{j=1}^{2J} z_j,$$

subject to the constraints $z \geq 0$, and $Bz = b$. Let z^* be the solution. We write

$$z^* = \begin{bmatrix} u^* \\ v^* \end{bmatrix}.$$

Then, as we shall see, $x^* = u^* - v^*$ minimizes the one-norm, subject to $Ax = b$.

First, we show that $u_j^* v_j^* = 0$, for each j. If, say, there is a j such that $0 < v_j^* \leq u_j^*$, then we can create a new vector z from z^* by replacing the old u_j^* with $u_j^* - v_j^*$ and the old v_j^* with zero, while maintaining $Bz = b$. But then, since $u_j^* - v_j^* < u_j^* + v_j^*$, it follows that $c^T z < c^T z^*$, which is a contradiction. Consequently, we have $\|x^*\|_1 = c^T z^*$.

Now we select any x with $Ax = b$. Write $u_j = x_j$, if $x_j \geq 0$, and $u_j = 0$, otherwise. Let $v_j = u_j - x_j$, so that $x = u - v$. Then let

$$z = \begin{bmatrix} u \\ v \end{bmatrix}.$$

Then $b = Ax = Bz$, and $c^T z = \|x\|_1$. Therefore

$$\|x^*\|_1 = c^T z^* \leq c^T z = \|x\|_1,$$

and x^* must be a minimum one-norm solution.

The reader is invited to provide an example showing that a minimum one-norm solution of $Ax = b$ need not be unique.

10.13.2 Image Reconstruction

In image reconstruction from limited linear-functional data, the vector x is nonnegative and arises as a vectorization of a two-dimensional image. The data we have pertaining to x is linear and takes the form $Ax = b$, for some matrix A and vector b. Typically, the problem is under-determined, since the number of entries of x is the number of pixels in the image, which we can make as large as we wish. The problem then is to select, from among all the feasible images, one particular one that has a good chance of being near the correct image. One approach is to take the solution of $Ax = b$ having the minimum Euclidean norm, $||x||_2$. Algorithms such as the projected ART and projected Landweber iterative methods can be used to find such solutions.

Another approach is to find the nonnegative solution of $Ax = b$ for which the one-norm,

$$||x||_1 = \sum_{j=1}^{J} |x_j|,$$

is minimized [98]. Since the x_j are to be nonnegative, the problem becomes the following: Minimize

$$f(x) = \sum_{j=1}^{J} x_j,$$

subject to

$$g_i(x) = (Ax)_i - b_i = 0,$$

for $i = 1, ..., I$, and

$$g_i(x) = -x_{i-I} \leq 0,$$

for $i = I + 1, ..., I + J$.

When the system $Ax = b$ is under-determined, the minimum one-norm solution tends to be sparser than the minimum two-norm solution. A simple example will illustrate this point.

Consider the equation $x + 2y = 1$. The minimum two-norm solution is $(0.2, 0.4)$, with two-norm $\frac{\sqrt{5}}{5}$, which is about 0.4472, but one-norm equal to 0.6. The solution $(0, 0.5)$ has two-norm and one-norm equal to 0.5, and the solution $(1.0, 0)$ has two-norm and one-norm equal to 1.0. Therefore, the minimum one-norm solution is $(0, 0.5)$, not $(0.2, 0.4)$.

We can write the one-norm of the vector x as

$$||x||_1 = \sum_{j=1}^{J} \frac{|x_j|^2}{|x_j|}.$$

The PDFT approach to image reconstruction [34, 35, 53, 64] selects the solution of $Ax = b$ that minimizes the weighted two-norm

$$||x||_w^2 = \sum_{j=1}^{J} \frac{|x_j|^2}{p_j} = \sum_{j=1}^{J} |x_j|^2 w_j,$$

where $p_j > 0$ is a prior estimate of the nonnegative image x to be reconstructed, and $w_j = p_j^{-1}$. To the extent that p_j accurately models the main features of x, such as which x_j are nearly zero and which are not, the two approaches should give similar reconstructions. The PDFT can be implemented using the ART algorithm (see [189, 190, 191]).

10.14 Exercises

Ex. 10.1 *Prove Proposition 10.1.*

Ex. 10.2 *Prove Corollary 10.1.*

Ex. 10.3 *Show that, although* $K(1,1) = 0$, *which is the saddle value, the point* $(1,1)$ *is not a saddle point for the function* $K(x,y) = x^2 - y^2$.

Ex. 10.4 *Prove Theorem 10.3.*

Ex. 10.5 *Apply the gradient form of the KKT Theorem to minimize the function* $f(x,y) = (x+1)^2 + y^2$ *over all* $x \geq 0$ *and* $y \geq 0$.

Ex. 10.6 [112] *Consider the following problem: Minimize the function*

$$f(x,y) = |x-2| + |y-2|,$$

subject to

$$g(x,y) = y^2 - x \leq 0,$$

and

$$h(x,y) = x^2 + y^2 - 1 = 0.$$

Illustrate this problem graphically, showing lines of constant value of f and the feasible region of points satisfying the constraints. Where is the solution of the problem? Where is the solution, if the equality constraint is removed? Where is the solution, if both constraints are removed?

Ex. 10.7 [176, Ex. 5.2.9 (a)] *Minimize the function*

$$f(x,y) = \sqrt{x^2 + y^2},$$

subject to

$$x + y \leq 0.$$

Show that the function $MP(z)$ *is not differentiable at* $z = 0$.

Ex. 10.8 [176, Ex. 5.2.9 (b)] *Minimize the function*

$$f(x,y) = -2x - y,$$

subject to

$$x + y \leq 1,$$
$$0 \leq x \leq 1,$$

and

$$y \geq 0.$$

Again, show that the function $MP(z)$ is not differentiable at $z = 0$.

Ex. 10.9 (Duffin [176, Ex. 5.2.9 (c)]) *Minimize the function*

$$f(x,y) = e^{-y},$$

subject to

$$\sqrt{x^2 + y^2} - x \leq 0.$$

Show that the function $MP(z)$ is not continuous at $z = 0$.

Ex. 10.10 *Apply the theory of convex programming to the primal Quadratic Programming Problem (QP), which is to minimize the function*

$$f(x) = \frac{1}{2}x^T Qx,$$

subject to

$$a^T x \leq c,$$

where $a \neq 0$ is in $\mathbb{R}^J$, $c < 0$ is real, and Q is symmetric, and positive-definite.

Ex. 10.11 *Use Theorem 10.6 to prove that any real N by N symmetric matrix has N mutually orthonormal eigenvectors.*

Chapter 11

Iterative Optimization

11.1 Chapter Summary

Now we begin our discussion of iterative methods for solving optimization problems. Topics include the role of the gradient operator, the Newton–Raphson (NR) method, and various computationally simpler variants of the NR method.

11.2 The Need for Iterative Methods

We know from beginning calculus that, if we want to optimize a differentiable function $g(x)$ of a single real variable x, we begin by finding the places where the derivative is zero, $g'(x) = 0$. Similarly, if we want to optimize a differentiable function $g(x)$ of a real vector variable x, we begin by finding the places where the gradient is zero, $\nabla g(x) = 0$. Generally, though, this is not the end of the story, for we still have to solve an equation for the optimal x. Unless we are fortunate, solving this equation algebraically may be computationally expensive, or may even be impossible, and we will need to turn to iterative methods. This suggests that we might use iterative methods to minimize $g(x)$ directly, and not solve an equation.

For example, suppose we wish to solve the over-determined system of linear equations $Ax = b$, but we don't know if the system has solutions. In that case, we may wish to minimize the function

$$g(x) = \frac{1}{2}\|Ax - b\|_2^2,$$

to get a least-squares solution. We know from linear algebra that if the matrix A^TA is invertible, then the unique minimizer of $g(x)$ is given by

$$x^* = (A^TA)^{-1}A^Tb.$$

In many applications, the number of equations and the number of unknowns may be quite large, making it expensive even to calculate the entries of the matrix A^TA. In such cases, we can find x^* using an iterative method such as Landweber's Algorithm, which has the iterative step

$$x^{k+1} = x^k + \gamma A^T(b - Ax^k).$$

The sequence $\{x^k\}$ converges to x^* for any value of γ in the interval $(0, 2/\lambda_{max})$, where λ_{max} is the largest eigenvalue of the matrix A^TA.

11.3 Optimizing Functions of a Single Real Variable

Suppose $g : \mathbb{R} \to \mathbb{R}$ is differentiable and attains its minimum value. We want to minimize the function $g(x)$. Solving $g'(x) = 0$ to find the optimal $x = x^*$ may not be easy, so we may turn to an iterative algorithm for finding roots of $g'(x)$, or one that minimizes $g(x)$ directly. Consider the following iterative procedure

$$x^{k+1} = x^k - \gamma_k g'(x^k), \tag{11.1}$$

for some sequence $\{\gamma_k\}$ of positive numbers. Such iterative procedures are called *descent algorithms* because, if $g'(x^k) > 0$, then we want to move to the left of x^k, while, if $g'(x^k) < 0$, we want to move to the right. If, at each step, we determine γ_k by minimizing the function of γ given by $f(x^k - \gamma g'(x^k))$, the iterative algorithm in Equation (11.1) is called the *steepest descent* method.

We shall be particularly interested in algorithms in which $\gamma_k = \gamma$ for all k. We denote by T the operator

$$Tx = x - \gamma g'(x).$$

Then, using $g'(x^*) = 0$, we find that

$$|x^* - x^{k+1}| = |Tx^* - Tx^k|.$$

11.4 Iteration and Operators

The iterative methods we shall consider involve the calculation of a sequence $\{x^k\}$ of vectors in $\mathbb{R}^J$, according to the formula $x^{k+1} = Tx^k$, where T is some function $T : \mathbb{R}^J \to \mathbb{R}^J$; such functions are called *operators* on $\mathbb{R}^J$. The operator $Tx = x - g'(x)$ above is an operator on $\mathbb{R}$.

Definition 11.1 *An operator T on $\mathbb{R}^J$ is* continuous *at x in the interior of its domain if*

$$\lim_{z \to x} \|Tz - Tx\| = 0.$$

The sequences generated by iterative methods can then be written $\{T^k x^0\}$, where $x = x^0$ is the starting point for the iteration and T^k means apply the operator T k times. All the operators we shall consider are continuous. If the sequence $\{x^k\}$ converges to a limit vector $\hat{x}$ in the domain of T, then, taking the limit, as $k \to +\infty$, on both sides of

$$x^{k+1} = Tx^k,$$

and using the continuity of the operator T, we have

$$\hat{x} = T\hat{x},$$

that is, the limit vector $\hat{x}$ is a *fixed point* of T.

Definition 11.2 *A vector x in the domain of the operator T is a* fixed point *of T if $T\hat{x} = \hat{x}$. The set of all fixed points of T is denoted* Fix(T).

We have several concerns, when we use iterative methods:

- Does the operator T have any fixed points?
- Does the sequence $\{T^k x^0\}$ converge?
- Does convergence depend on the choice of x^0?
- If it does converge, does the limit depend on the starting vector x^0? If so, how?
- When the sequence $\{T^k x^0\}$ converges, is the limit a solution to our problem?
- How fast does the sequence $\{T^k x^0\}$ converge?
- How difficult is it to perform a single step, going from x^k to x^{k+1}?

To answer these questions, we will need to learn about the properties of the particular operator T being used.

We begin our study of iterative optimization algorithms with the well-known Newton–Raphson method and its various modifications. Then we turn to gradient descent methods, particularly as they apply to convex functions.

11.5 The Newton–Raphson Approach

The Newton–Raphson approach to minimizing a real-valued function $f : \mathbb{R}^J \to \mathbb{R}$ involves finding x^* such that $\nabla f(x^*) = 0$.

11.5.1 Functions of a Single Variable

We begin with the problem of finding a root of a function $g : \mathbb{R} \to \mathbb{R}$. If x^0 is not a root, compute the line tangent to the graph of g at $x = x^0$ and let x^1 be the point at which this line intersects the horizontal axis; that is,

$$x^1 = x^0 - g(x^0)/g'(x^0).$$

Continuing in this fashion, we have

$$x^{k+1} = x^k - g(x^k)/g'(x^k).$$

This is the *Newton–Raphson algorithm* for finding roots. Convergence, when it occurs, is usually more rapid than steepest descent, but requires that x^0 be sufficiently close to the solution.

Now suppose that $f : \mathbb{R} \to \mathbb{R}$ is a real-valued function that we wish to minimize by solving $f'(x) = 0$. Letting $g(x) = f'(x)$ and applying the Newton–Raphson algorithm to $g(x)$ gives the iterative step

$$x^{k+1} = x^k - f'(x^k)/f''(x^k).$$

This is the Newton–Raphson optimization algorithm. Now we extend these results to functions of several variables.

11.5.2 Functions of Several Variables

The Newton–Raphson algorithm for finding roots of functions $g : \mathbb{R}^J \to \mathbb{R}^J$ has the iterative step

$$x^{k+1} = x^k - [\mathcal{J}(g)(x^k)]^{-1} g(x^k),$$

where $\mathcal{J}(g)(x)$ is the Jacobian matrix of first partial derivatives, $\frac{\partial g_m}{\partial x_j}(x^k)$, for $g(x) = (g_1(x), ..., g_J(x))^T$.

To minimize a function $f : \mathbb{R}^J \to \mathbb{R}$, we let $g(x) = \nabla f(x)$ and find a root of g. Then the Newton–Raphson iterative step becomes

$$x^{k+1} = x^k - [\nabla^2 f(x^k)]^{-1} \nabla f(x^k), \tag{11.2}$$

where $\nabla^2 f(x) = \mathcal{J}(g)(x)$ is the Hessian matrix of second partial derivatives of f.

The quadratic approximation to $f(x)$ around the point x^k is

$$f(x) \approx f(x^k) + \langle \nabla f(x^k), x - x^k \rangle + \frac{1}{2}(x - x^k)^T \nabla^2 f(x^k)(x - x^k).$$

The right side of this equation attains its minimum value when

$$0 = \nabla f(x^k) + \nabla^2 f(x^k)(x - x^k),$$

that is, when $x = x^{k+1}$ as given by Equation (11.2). If $f(x)$ is a quadratic function, that is,

$$f(x) = x^T Qx + x^T b + c,$$

for constant invertible matrix Q and constant vectors b and c, then the Newton–Raphson iteration converges to the answer in one step. Therefore, if $f(x)$ is close to quadratic, the convergence should be reasonably rapid. This leads to the notion of *self-concordant functions*, for which the third derivative of $f(x)$ is small, relative to the second derivative [164].

From the quadratic approximation

$$f(x^{k+1}) \approx f(x^k) + \langle \nabla f(x^k), x^{k+1} - x^k \rangle + \frac{1}{2}(x^{k+1} - x^k)^T \nabla^2 f(x^k)(x^{k+1} - x^k),$$

and the formula for the iterative NR step we find that

$$f(x^{k+1}) - f(x^k) \approx -\frac{1}{2} \nabla f(x^k)^T [\nabla^2 f(x^k)]^{-1} \nabla f(x^k).$$

If the Hessian matrix $\nabla^2 f(x^k)$ is always positive-definite, which may not be the case, then its inverse will also be positive-definite and the NR step will reduce the value of the objective function $f(x)$. One area of research in the intersection of numerical linear algebra and optimization focuses on finding positive-definite approximations of the Hessian matrix [203].

11.6 Approximate Newton–Raphson Methods

To use the NR method to minimize $f(x)$, at each step of the iteration we need to solve a system of equations involving the Hessian matrix for f. There are many iterative procedures designed to retain much of the advantages of the NR method, while avoiding the use of the Hessian matrix, or, indeed, while avoiding the use of the gradient. These methods are discussed in most texts on numerical methods [164]. We sketch briefly some of these approaches.

11.6.1 Avoiding the Hessian Matrix

Quasi-Newton methods, designed to avoid having to calculate the Hessian matrix, are often used instead of the Newton–Raphson algorithm. The iterative step of the quasi-Newton methods is

$$x^{k+1} = x^k - B_k^{-1} \nabla f(x^k),$$

where the matrix B_k is an approximation of $\nabla^2 f(x^k)$ that is easier to compute.

In the case of $g : \mathbb{R} \to \mathbb{R}$, the second derivative of $g(x)$ is approximately

$$g''(x^k) \approx \frac{g'(x^k) - g'(x^{k-1})}{x^k - x^{k-1}}.$$

This suggests that, for the case of functions of several variables, the matrix B_k should be selected so that

$$B_k(x^k - x^{k-1}) = \nabla f(x^k) - \nabla f(x^{k-1}). \tag{11.3}$$

In addition to satisfying Equation (11.3), the matrix B_k should also be symmetric and positive-definite. Finally, we should be able to obtain B_{k+1} relatively easily from B_k.

11.6.2 The BFGS Method

The Broyden, Fletcher, Goldfarb, and Shanno (BFGS) method uses the rank-two update formula

$$B_{k+1} = B_k - \frac{(B_k s^k)(B_k s^k)^T}{(s^k)^T B_k s^k} + \frac{y^k (y^k)^T}{(y^k)^T s^k},$$

with

$$s^k = x^{k+1} - x^k,$$

and

$$y^k = \nabla f(x^{k+1}) - \nabla f(x^k).$$

11.6.3 The Broyden Class

A general class of update methods, known as the Broyden class, uses the update formula

$$B_{k+1} = B_k - \frac{(B_k s^k)(B_k s^k)^T}{(s^k)^T B_k s^k} + \frac{y^k (y^k)^T}{(y^k)^T s^k} + \phi((s^k)^T B_k s^k) u^k (u^k)^T,$$

with ϕ a scalar and

$$u^k = \frac{y^k}{(y^k)^T s^k} - \frac{B_k s^k}{(s^k)^T B_k s^k}.$$

When $\phi = 0$ we get the BFGS method, while the choice of $\phi = 1$ gives the Davidon, Fletcher, and Powell (DFP) method.

Note that for the updates in the Broyden class, the matrix B_{k+1} has the form

$$B_{k+1} = B_k + a^k(a^k)^T + b^k(b^k)^T + c^k(c^k)^T,$$

for certain vectors a^k, b^k and c^k. Therefore, the inverse of B_{k+1} can be obtained easily from the inverse of B_k, with three applications of the Sherman–Morrison–Woodbury Identity (see Exercise 6.4).

11.6.4 Avoiding the Gradient

Quasi-Newton methods use an approximation of the Hessian matrix that is simpler to calculate, but still employ the gradient at each step. For functions $g : \mathbb{R} \to \mathbb{R}$, the derivative can be approximated by a *finite difference*, that is,

$$g'(x^k) \approx \frac{g(x^k) - g(x^{k-1})}{x^k - x^{k-1}}.$$

In the case of functions of several variables, the gradient vector can be approximated by using a finite-difference approximation for each of the first partial derivatives.

11.7 Derivative-Free Methods

In many important applications, calculating values of the function to be optimized is expensive and calculating gradients impractical. In such cases, it is common to use *direct-search methods.* Generally, these are iterative methods that are easy to program, do not employ derivatives or their approximations, require relatively few function evaluations, and are useful even when the measurements are noisy.

11.7.1 Multi-Directional Search Algorithms

Methods such as the *multi-directional search* algorithms begin with the values of the function $f(x)$ at $J+1$ points, where x is in $\mathbb{R}^J$, and then use these values to move to a new set of points. These points are chosen to describe a simplex pattern in $\mathbb{R}^J$, that is, they do not all lie on a single hyperplane in $\mathbb{R}^J$. For that reason, these methods are sometimes called *simplex* methods, although they are unrelated to Dantzig's method of the same name. The Nelder–Mead algorithm [165, 143, 156] is one such simplex algorithm.

11.7.2 The Nelder–Mead Algorithm

For simplicity, we follow McKinnon [156] and describe the Nelder–Mead (NM) algorithm only for the case of $J = 2$. The NM algorithm begins with the choice of vertices:

ORDER: Obtain b, s, and w, with

$$f(b) \leq f(s) \leq f(w).$$

Then take

$$m = \frac{1}{2}(b + s).$$

Let the *search line* be

$$L(\rho) = m + \rho(m - w),$$

and

$$r = L(1) = 2m - w.$$

- **{if** $f(r) < f(b)$**}** let $e = L(2)$. If $f(e) < f(b)$ *accept* e; otherwise *accept* r.
- **{if** $f(b) \leq f(r)$**}** then
  - **{if** $f(r) < f(s)$**}** *accept* r.
  - **{if** $f(s) \leq f(r)$**}**
    - **{if** $f(r) < f(w)$**}** let $c = L(0.5)$
      - **{if** $f(c) \leq f(r)$**}** *accept* c;
      - **{if** $f(r) < f(c)$**}** go to SHRINK.
    - **{if** $f(w) \leq f(r)$**}** let $c = L(-0.5)$.
      - **{if** $f(c) < f(w)$**}** *accept* c; otherwise go to SHRINK.

Replace w with the *accepted* point and go to ORDER.
SHRINK: Replace s with $\frac{1}{2}(s + b)$ and w with $\frac{1}{2}(w + b)$; go to ORDER.

11.7.3 Comments on the Nelder–Mead Algorithm

Although the Nelder–Mead algorithm is quite popular in many areas of applications, relatively little of a theoretical nature is known. The interested reader is directed to the papers [143, 156], as well as to more recent work by Margaret Wright of NYU. A good treatment of the Nelder–Mead algorithm, along with a number of other derivative-free techniques, is the new book by Conn, Scheinberg and Vicente [89].

11.8 Rates of Convergence

In this section we illustrate the concept of *rate of convergence* [30] by considering the fixed-point iteration $x_{k+1} = g(x_k)$, for the twice continuously differentiable function $g : \mathbb{R} \to \mathbb{R}$. We suppose that $g(z) = z$ and we are interested in the distance $|x_k - z|$.

11.8.1 Basic Definitions

Definition 11.3 *Suppose the sequence $\{x_k\}$ converges to z. If there are positive constants λ and α such that*

$$\lim_{k\to\infty} \frac{|x_{k+1} - z|}{|x_k - z|^\alpha} = \lambda,$$

then $\{x_k\}$ is said to converge to z with order α *and* asymptotic error constant λ. *If $\alpha = 1$, the convergence is said to be* linear*; if $\alpha = 2$, the convergence is said to be* quadratic.

11.8.2 Illustrating Quadratic Convergence

According to the Extended Mean Value Theorem,

$$g(x) = g(z) + g'(z)(x - z) + \frac{1}{2}g''(c)(x - z)^2,$$

for some c between x and z. Suppose now that $x_k \to z$ and, in addition, $g'(z) = 0$. Then we have

$$x_{k+1} = g(x_k) = z + \frac{1}{2}g''(c_k)(x_k - z)^2,$$

for some c_k between x_k and z. Therefore,

$$|x_{k+1} - z| = \frac{1}{2}|g''(c_k)|\,|x_k - z|^2,$$

and the convergence is quadratic, with $\lambda = |g''(z)|$.

11.8.3 Motivating the Newton–Raphson Method

Suppose that we are seeking a root z of the function $f : \mathbb{R} \to \mathbb{R}$. We define

$$g(x) = x - h(x)f(x),$$

for some function $h(x)$ to be determined. Then $f(z) = 0$ implies that $g(z) = z$. In order to have quadratic convergence of the iterative sequence $x_{k+1} = g(x_k)$, we want $g'(z) = 0$. From

$$g'(x) = 1 - h'(x)f(x) - h(x)f'(x),$$

it follows that we want

$$h(z) = 1/f'(z).$$

Therefore, we choose

$$h(x) = 1/f'(x),$$

so that

$$g(x) = x - f(x)/f'(x).$$

The iteration then takes the form

$$x_{k+1} = g(x_k) = x_k - f(x_k)/f'(x_k),$$

which is the Newton–Raphson iteration.

11.9 Descent Methods

Suppose that $g(x)$ is convex and the function $f(x) = g'(x)$ is L-Lipschitz. If $g(x)$ is twice differentiable, this would be the case if

$$0 \leq g''(x) \leq L,$$

for all x. If γ is in the interval $(0, \frac{2}{L})$, then the operator $Tx = x - \gamma g'(x)$ is an averaged operator; from the KMO Theorem 14.2, we know that the iterative sequence $\{T^k x^0\}$ converges to a minimizer of $g(x)$, whenever a minimizer exists.

If $g(x)$ is convex and $f(x) = g'(x)$ is L-Lipschitz, then $\frac{1}{L}g'(x)$ is non-expansive, so that, by Theorem 9.20 $\frac{1}{L}g'(x)$ is fne. Then, as we shall see in Chapter 14, the operator

$$Tx = x - \gamma g'(x)$$

is such that, whenever $0 < \gamma < \frac{2}{L}$, the iterative sequence $x^{k+1} = Tx^k = x^k - \gamma g'(x^k)$ converges to a minimizer of $g(x)$, whenever minimizers exist.

In the next section we extend these results to functions of several variables.

11.10 Optimizing Functions of Several Real Variables

Suppose $g : \mathbb{R}^J \to \mathbb{R}$ is differentiable and attains its minimum value. We want to minimize the function $g(x)$. Solving $\nabla g(x) = 0$ to find the optimal $x = x^*$ may not be easy, so we may turn to an iterative algorithm for finding roots of $\nabla g(x)$, or one that minimizes $g(x)$ directly. From Cauchy's Inequality, we know that the directional derivative of $g(x)$, at $x = a$, and in the direction of the vector unit vector d, satisfies

$$|g'(a; d)| = |\langle \nabla g(a), d \rangle| \leq \|\nabla g(a)\|_2 \, \|d\|_2,$$

and that $g'(a; d)$ attains its most positive value when the direction d is a positive multiple of $\nabla g(a)$. This suggests *steepest descent* optimization.

Steepest descent iterative optimization makes use of the fact that the direction of greatest increase of $g(x)$ away from $x = x^k$ is in the direction $d = \nabla g(x^k)$. Therefore, we select as the next vector in the iterative sequence

$$x^{k+1} = x^k - \gamma_k \nabla g(x^k),$$

for some $\gamma_k > 0$. Ideally, we would choose γ_k optimally, so that

$$g(x^k - \gamma_k \nabla g(x^k)) \leq g(x^k - \gamma \nabla g(x^k)), \tag{11.4}$$

for all $\gamma \geq 0$; that is, we would proceed away from x^k, in the direction of $-\nabla g(x^k)$, stopping just as $g(x)$ begins to increase. Then we call this point x^{k+1} and repeat the process.

Lemma 11.1 *Suppose that x^{k+1} is chosen using the optimal value of γ_k, as described by Equation (11.4). Then*

$$\langle \nabla g(x^{k+1}), \nabla g(x^k) \rangle = 0.$$

In practice, finding the optimal γ_k is not a simple matter. Instead, one can try a few values of α and accept the best of these few, or one can try to find a constant value γ of the parameter having the property that the iterative step

$$x^{k+1} = x^k - \gamma \nabla g(x^k)$$

leads to a convergent sequence. It is this latter approach that we shall consider here.

We denote by T the operator

$$Tx = x - \gamma \nabla g(x).$$

Then, using $\nabla g(x^*) = 0$, we find that

$$\|x^* - x^{k+1}\|_2 = \|Tx^* - Tx^k\|_2.$$

We would like to know if there are choices for γ that imply convergence of the iterative sequence. As in the case of functions of a single variable, for functions $g(x)$ that are *convex*, the answer is yes.

If $g(x)$ is convex and $F(x) = \nabla g(x)$ is L-Lipschitz, then $G(x) = \frac{1}{L}\nabla g(x)$ is firmly nonexpansive. Then, as we shall see in Chapter 14, for $\gamma > 0$, the operator

$$Tx = x - \gamma \nabla g(x)$$

is such that, whenever $0 < \gamma < \frac{2}{L}$, the iterative sequence $x^{k+1} = Tx^k = x^k - \gamma \nabla g(x^k)$ converges to a minimizer of $g(x)$, whenever minimizers exist.

For example, the function $g(x) = \frac{1}{2}\|Ax - b\|_2^2$ is convex and its gradient is

$$f(x) = \nabla g(x) = A^T(Ax - b).$$

A steepest descent algorithm for minimizing $g(x)$ then has the iterative step

$$x^{k+1} = x^k - \gamma_k A^T(Ax^k - b),$$

where the parameter γ_k should be selected so that

$$g(x^{k+1}) < g(x^k).$$

The linear operator that transforms each vector x into $A^T Ax$ has the property that

$$\|A^T Ax - A^T Ay\|_2 \leq \lambda_{max}\|x - y\|_2,$$

where λ_{max} is the largest eigenvalue of the matrix $A^T A$; this operator is then L-Lipschitz, for $L = \lambda_{max}$. Consequently, the operator that transforms x into $\frac{1}{L}A^T Ax$ is nonexpansive.

11.11 Projected Gradient-Descent Methods

As we have remarked previously, one of the fundamental problems in continuous optimization is to find a minimizer of a function over a subset of $\mathbb{R}^J$. The following propositions will help to motivate the projected gradient-descent algorithm.

Proposition 11.1 *Let $f : \mathbb{R}^J \to \mathbb{R}$ be convex and differentiable and let $C \subseteq \mathbb{R}^J$ be closed, nonempty and convex. Then $x \in C$ minimizes f over C if and only if*

$$\langle \nabla f(x), c - x \rangle \geq 0,$$

for all $c \in C$.

Proof: Since f is convex, we know from Theorem 9.16 that

$$f(b) - f(a) \geq \langle \nabla f(a), b - a \rangle,$$

for all a and b. Therefore, if

$$\langle \nabla f(x), c - x \rangle \geq 0,$$

for all $c \in C$, then $f(c) - f(x) \geq 0$ for all $c \in C$ also.

Conversely, suppose that $f(c) - f(x) \geq 0$, for all $c \in C$. For each $c \in C$, let $d = \frac{c-x}{\|c-x\|_2}$, so that

$$\langle \nabla f(x), d \rangle = \frac{1}{\|c - x\|_2} \langle \nabla f(x), c - x \rangle$$

is the directional derivative of f at x, in the direction of c. Because $f(c) - f(x) \geq 0$, for all $c \in C$, this directional derivative must be nonnegative. ∎

Proposition 11.2 *Let $f : \mathbb{R}^J \to \mathbb{R}$ be convex and differentiable and let $C \subseteq \mathbb{R}^J$ be closed, nonempty and convex. Then $x \in C$ minimizes f over C if and only if*

$$x = P_C(x - \gamma \nabla f(x)),$$

for all $\gamma > 0$.

Proof: By Proposition 4.4, we know that $x = P_C(x - \gamma \nabla f(x))$ if and only if

$$\langle x - (x - \gamma \nabla f(x)), c - x \rangle \geq 0,$$

for all $c \in C$. But this is equivalent to

$$\langle \nabla f(x), c - x \rangle \geq 0,$$

for all $c \in C$, which, by the previous proposition, is equivalent to x minimizing the function f over all $c \in C$. ∎

This leads us to the projected gradient-descent algorithm. According to the previous proposition, we know that x minimizes f over C if and only if x is a fixed point of the operator

$$Tx = P_C(x - \gamma \nabla f(x)).$$

In the next section we present an elementary proof of the following theorem.

Theorem 11.1 *Let $f : \mathbb{R}^J \to \mathbb{R}$ be convex and differentiable, with ∇f L-Lipschitz. Let C be any closed, convex subset of $\mathbb{R}^J$. For $0 < \gamma < \frac{1}{L}$, let $T = P_C(I - \gamma \nabla f)$. If T has fixed points, then the sequence $\{x^k\}$ given by $x^{k+1} = Tx^k$ converges to a fixed point of T, which is then a minimizer of f over C.*

The iterative step is given by

$$x^{k+1} = P_C(x^k - \gamma \nabla f(x^k)). \tag{11.5}$$

Any fixed point of the operator T minimizes the function $f(x)$ over x in C.

This theorem is a corollary of the KMO Theorem 14.2 for *averaged* operators, which we shall define in Chapter 14. Using the KMO Theorem it can be shown that convergence holds for $0 < \gamma < \frac{2}{L}$. The proof given in the next section employs auxiliary-function methods and avoids using the non-trivial results that, because the operator $\frac{1}{L}\nabla f$ is nonexpansive, it is firmly nonexpansive (see Theorem 9.20), and that the product of averaged operators is again averaged (see Proposition 14.1).

11.12 Auxiliary-Function Methods

In this section we introduce the notion of *auxiliary-function* methods, a topic we shall consider in more detail in Chapter 15. The problem is to minimize a function $f : X \to \mathbb{R}$, over a subset $C \subseteq X$, where X is an arbitrary set. At the kth step of the iteration we minimize a function

$$G_k(x) = f(x) + g_k(x),$$

to obtain x^k. The auxiliary functions $g_k(x)$ are selected to enforce the constraint that x be in C, as in barrier-function methods, or to penalize violations of that constraint, such as in penalty-function methods.

In AF methods certain restrictions are placed on the auxiliary functions $g_k(x)$ to control the behavior of the sequence $\{f(x^k)\}$. Even when there are no constraints, the problem of minimizing a real-valued function may require iteration; the formalism of AF minimization can be useful in deriving such iterative algorithms, as well as in proving convergence, as we shall see in the next section. As originally formulated, barrier- and penalty-function algorithms are not in the AF class, but can be reformulated as AF algorithms. In AF methods the auxiliary functions satisfy additional properties that guarantee that the sequence $\{f(x^k)\}$ is nonincreasing.

We can use auxiliary-function (AF) methods to prove Theorem 11.1. For each $k = 1, 2, ...$ let

$$G_k(x) = f(x) + \frac{1}{2\gamma}\|x - x^{k-1}\|_2^2 - D_f(x, x^{k-1}),$$

where

$$D_f(x, x^{k-1}) = f(x) - f(x^{k-1}) - \langle \nabla f(x^{k-1}), x - x^{k-1} \rangle.$$

Since $f(x)$ is convex, $D_f(x,y) \geq 0$ for all x and y and is the Bregman distance formed from the function f [25].

The auxiliary function

$$g_k(x) = \frac{1}{2\gamma}\|x - x^{k-1}\|_2^2 - D_f(x, x^{k-1})$$

can be rewritten as

$$g_k(x) = D_h(x, x^{k-1}),$$

where

$$h(x) = \frac{1}{2\gamma}\|x\|_2^2 - f(x).$$

Therefore, $g_k(x) \geq 0$ whenever $h(x)$ is a convex function.

We know that $h(x)$ is convex if and only if

$$\langle \nabla h(x) - \nabla h(y), x - y\rangle \geq 0,$$

for all x and y. This is equivalent to

$$\frac{1}{\gamma}\|x - y\|_2^2 - \langle \nabla f(x) - \nabla f(y), x - y\rangle \geq 0. \tag{11.6}$$

Since ∇f is L-Lipschitz, the inequality (11.6) holds whenever $0 < \gamma < \frac{1}{L}$.

Lemma 11.2 *The x^k that minimizes $G_k(x)$ over $x \in C$ is given by Equation (11.5).*

Proof: We know that

$$\langle \nabla G_k(x^k), x - x^k\rangle \geq 0,$$

for all $x \in C$. With

$$\nabla G_k(x^k) = \frac{1}{\gamma}(x^k - x^{k-1}) + \nabla f(x^{k-1}),$$

we have

$$\langle x^k - (x^{k-1} - \gamma\nabla f(x^{k-1})), x - x^k\rangle \geq 0,$$

for all $x \in C$. We then conclude that

$$x^k = P_C(x^{k-1} - \gamma\nabla f(x^{k-1})).$$

∎

A relatively simple calculation shows that

$$G_k(x) - G_k(x^k) = \frac{1}{2\gamma}\|x - x^k\|_2^2 + \frac{1}{\gamma}\langle x^k - (x^{k-1} - \gamma\nabla f(x^{k-1})), x - x^k\rangle. \quad (11.7)$$

From Equation (11.5) it follows that

$$G_k(x) - G_k(x^k) \geq \frac{1}{2\gamma}\|x - x^k\|_2^2, \quad (11.8)$$

for all $x \in C$, so that

$$G_k(x) - G_k(x^k) \geq \frac{1}{2\gamma}\|x - x^k\|_2^2 - D_f(x, x^k) = g_{k+1}(x).$$

Now let $\hat{x}$ minimize $f(x)$ over all $x \in C$. Then

$$G_k(\hat{x}) - G_k(x^k) = f(\hat{x}) + g_k(\hat{x}) - f(x^k) - g_k(x^k)$$
$$\leq f(\hat{x}) + G_{k-1}(\hat{x}) - G_{k-1}(x^{k-1}) - f(x^k) - g_k(x^k),$$

so that

$$\Big(G_{k-1}(\hat{x}) - G_{k-1}(x^{k-1})\Big) - \Big(G_k(\hat{x}) - G_k(x^k)\Big) \geq f(x^k) - f(\hat{x}) + g_k(x^k) \geq 0.$$

Therefore, the sequence $\{G_k(\hat{x}) - G_k(x^k)\}$ is decreasing and the sequences $\{g_k(x^k)\}$ and $\{f(x^k) - f(\hat{x})\}$ converge to zero.

From

$$G_k(\hat{x}) - G_k(x^k) \geq \frac{1}{2\gamma}\|\hat{x} - x^k\|_2^2,$$

it follows that the sequence $\{x^k\}$ is bounded. Let $\{x^{k_n}\}$ converge to $x^* \in C$ with $\{x^{k_n+1}\}$ converging to $x^{**} \in C$; we then have $f(x^*) = f(x^{**}) = f(\hat{x})$.

Replacing the generic $\hat{x}$ with x^{**}, we find that $\{G_{k_n+1}(x^{**}) - G_{k_n+1}(x^{k_n+1})\}$ is decreasing. By Equation (11.7), this subsequence converges to zero; therefore, the entire sequence $\{G_k(x^{**}) - G_k(x^k)\}$ converges to zero. From the inequality in (11.8), we conclude that the sequence $\{\|x^{**} - x^k\|_2^2\}$ converges to zero, and so $\{x^k\}$ converges to x^{**}. This completes the proof of Theorem 11.1.

11.13 Feasible-Point Methods

We consider now the problem of minimizing a function $f(x) : \mathbb{R}^J \to \mathbb{R}$, subject to the equality constraints $Ax = b$, where A is an I by J real matrix, with rank I and $I < J$. We assume that the gradient ∇f is L-Lipschitz continuous. The methods we consider here are *feasible-point methods*, also called *interior-point methods.*

11.13.1 The Projected Gradient Algorithm

Let C be the set of all x in $\mathbb{R}^J$ such that $Ax = b$. Let $\hat{x}$ be an arbitrary member of C. Then every point x in C can be written as $x = w + \hat{x}$, for some w in $NS(A)$, the null space of A. For simplicity, we take for $\hat{x}$ the minimum-norm solution, $\hat{x} = A^T(AA^T)^{-1}b$. Then we have

$$C = NS(A) + A^T(AA^T)^{-1}b,$$

and, for every z in $\mathbb{R}^J$, we have

$$P_C z = P_{NS(A)} z + A^T(AA^T)^{-1}b.$$

Using

$$P_{NS(A)} z = z - A^T(AA^T)^{-1}Az,$$

we have

$$P_C z = z + A^T(AA^T)^{-1}(b - Az).$$

Now the iteration in Equation (11.5) becomes

$$c^{k+1} = c^k - \gamma P_{NS(A)} \nabla f(c^k);$$

we use c^k instead of x^k to remind us that each iterate lies in the set C. The sequence $\{c^k\}$ converges to a solution for any γ in $(0, \frac{1}{L})$, whenever solutions exist. We call this method the *projected gradient algorithm.*

In the next subsection we present a somewhat simpler approach.

11.13.2 Reduced Gradient Methods

Let c^0 be a *feasible point,* that is, $Ac^0 = b$. Then $c = c^0 + p$ is also feasible if p is in the null space of A, that is, $Ap = 0$. Let Z be a J by $J - I$ matrix whose columns form a basis for the null space of A. We want $p = Zv$ for some v. The best v will be the one for which the function

$$\phi(v) = f(c^0 + Zv)$$

is minimized. We can apply to the function $\phi(v)$ the steepest descent method, or Newton–Raphson or any other minimization technique.

The steepest descent method, applied to $\phi(v)$, is called the *reduced steepest descent method* [164]. The gradient of $\phi(v)$, also called the *reduced gradient*, is

$$\nabla\phi(v) = Z^T\nabla f(c),$$

where $c = c^0 + Zv$. We choose the matrix Z so that $\rho(Z^TZ) \leq 1$, so that the gradient operator $\nabla\phi$ is L-Lipschitz.

For the *reduced gradient algorithm*, the iteration in Equation (11.5) becomes

$$v^{k+1} = v^k - \gamma\nabla\phi(v^k),$$

so that the iteration for $c^{k+1} = c^0 + Zv^{k+1}$ is

$$c^{k+1} = c^k - \gamma ZZ^T\nabla f(c^k).$$

The vectors c^k are feasible, that is, lie in C, and the sequence $\{c^k\}$ converges to a solution, whenever solutions exist, for any $0 < \gamma < \frac{1}{L}$.

11.13.3 The Reduced Newton–Raphson Method

The next method we consider is a modification of the Newton–Raphson method, in which we begin with a feasible point and each NR step is in the null space of the matrix A, to maintain the condition $Ax = b$. The discussion here is taken from [164].

Once again, our objective is to minimize $\phi(v)$. The Newton–Raphson method, applied to $\phi(v)$, is called the *reduced Newton–Raphson method.* The Hessian matrix of $\phi(v)$, also called the *reduced Hessian matrix*, is

$$\nabla^2\phi(v) = Z^T\nabla^2 f(x)Z,$$

where $x = \hat{x} + Zv$, so algorithms to minimize $\phi(v)$ can be written in terms of the gradient and Hessian of f itself.

The reduced NR algorithm can then be viewed in terms of the vectors $\{v^k\}$, with $v^0 = 0$ and

$$v^{k+1} = v^k - [\nabla^2\phi(v^k)]^{-1}\nabla\phi(v^k);$$

the corresponding x^k is

$$x^k = \hat{x} + Zv^k.$$

11.13.4 An Example

Consider the problem of minimizing the function

$$f(x) = \frac{1}{2}x_1^2 - \frac{1}{2}x_3^2 + 4x_1x_2 + 3x_1x_3 - 2x_2x_3,$$

subject to

$$x_1 - x_2 - x_3 = -1.$$

Let $\hat{x} = [1, 1, 1]^T$. Then the matrix A is $A = [1, -1, -1]$ and the vector b is $b = [-1]$. Let the matrix Z be

$$Z = \begin{bmatrix} 1 & 1 \\ 1 & 0 \\ 0 & 1 \end{bmatrix}.$$

The reduced gradient at $\hat{x}$ is then

$$Z^T \nabla f(\hat{x}) = \begin{bmatrix} 1 & 1 & 0 \\ 1 & 0 & 1 \end{bmatrix} \begin{bmatrix} 8 \\ 2 \\ 0 \end{bmatrix} = \begin{bmatrix} 10 \\ 8 \end{bmatrix},$$

and the reduced Hessian matrix at $\hat{x}$ is

$$Z^T \nabla^2 f(\hat{x}) Z = \begin{bmatrix} 1 & 1 & 0 \\ 1 & 0 & 1 \end{bmatrix} \begin{bmatrix} 1 & 4 & 3 \\ 4 & 0 & -2 \\ 3 & -2 & -1 \end{bmatrix} \begin{bmatrix} 1 & 1 \\ 1 & 0 \\ 0 & 1 \end{bmatrix} = \begin{bmatrix} 9 & 6 \\ 6 & 6 \end{bmatrix}.$$

Then the reduced Newton–Raphson equation yields

$$v = \begin{bmatrix} -2/3 \\ -2/3 \end{bmatrix},$$

and the reduced Newton–Raphson direction is

$$p = Zv = \begin{bmatrix} -4/3 \\ -2/3 \\ -2/3 \end{bmatrix}.$$

Since the function $\phi(v)$ is quadratic, one reduced Newton–Raphson step suffices to obtain the solution, $x^* = [-1/3, 1/3, 1/3]^T$.

11.13.5 A Primal-Dual Approach

Once again, the objective is to minimize the function $f(x) : \mathbb{R}^J \to \mathbb{R}$, subject to the equality constraints $Ax = b$. According to the Karush–Kuhn–Tucker Theorem 10.5, $\nabla L(x, \lambda) = 0$ at the optimal values of x and λ, where the Lagrangian $L(x, \lambda)$ is

$$L(x, \lambda) = f(x) + \lambda^T (b - Ax).$$

Finding a zero of the gradient of $L(x, \lambda)$ means that we have to solve the equations

$$\nabla f(x) - A^T \lambda = 0$$

and

$$Ax = b.$$

We define the function $G(x, \lambda)$ taking values in $\mathbb{R}^J \times \mathbb{R}^I$ to be

$$G(x, \lambda) = (\nabla f(x) - A^T \lambda, Ax - b)^T.$$

We then apply the NR method to find a zero of the function G. The Jacobian matrix for G is

$$J_G(x, \lambda) = \begin{bmatrix} \nabla^2 f(x) & -A^T \\ A & 0 \end{bmatrix},$$

so one step of the NR method is

$$(x^{k+1}, \lambda^{k+1})^T = (x^k, \lambda^k)^T - J_G(x^k, \lambda^k)^{-1} G(x^k, \lambda^k).$$

We can rewrite this as

$$\nabla^2 f(x^k)(x^{k+1} - x^k) - A^T(\lambda^{k+1} - \lambda^k) = A^T \lambda^k - \nabla f(x^k),$$

and

$$A(x^{k+1} - x^k) = b - Ax^k. \tag{11.9}$$

It follows from Equation (11.9) that $Ax^{k+1} = b$, for $k = 0, 1, \dots$, so that this primal-dual algorithm is a feasible-point algorithm.

11.14 Quadratic Programming

The *quadratic-programming* problem (QP) is to minimize a quadratic function, subject to inequality constraints and, often, the nonnegativity of the variables. Using the Karush–Kuhn–Tucker Theorem 10.6 for mixed constraints and introducing slack variables, this problem can be reformulated as a linear programming problem and solved by Wolfe's Algorithm [176], a variant of the simplex method. In the case of general constrained optimization, the Newton–Raphson method for finding a stationary point of the Lagrangian can be viewed as solving a sequence of quadratic programming problems. This leads to *sequential quadratic programming* [164].

11.14.1 The Quadratic-Programming Problem

The primal QP problem is to minimize the quadratic function

$$f(x) = a + x^T c + \frac{1}{2} x^T Q x,$$

subject to the constraints

$$Ax \leq b,$$

and $x_j \geq 0$, for $j = 1, ..., J$. Here a, b, and c are given, Q is a given J by J positive-definite matrix with entries Q_{ij}, and A is an I by J matrix with rank I and entries A_{ij}. To allow for some equality constraints, we say that

$$(Ax)_i \leq b_i,$$

for $i = 1, ..., K$, and

$$(Ax)_i = b_i,$$

for $i = K + 1, ..., I$.

We incorporate the nonnegativity constraints $x_j \geq 0$ by requiring

$$-x_j \leq 0,$$

for $j = 1, ..., J$. Applying the Karush–Kuhn–Tucker Theorem to this problem, we find that if a regular point x^* is a solution, then there are vectors μ^* and ν^* such that

(1) $\mu_i^* \geq 0$, for $i = 1, ..., K$;

(2) $\nu_j^* \geq 0$, for $j = 1, ..., J$;

(3) $c + Qx^* + A^T\mu^* - v^* = 0$;

(4) $\mu_i^*((Ax^*)_i - b_i) = 0$, for $i = 1, ..., I$;

(5) $x_j^*\nu_j^* = 0$, for $j = 1, ..., J$.

One way to solve this problem is to reformulate it as a linear-programming problem. To that end, we introduce slack variables x_{J+i}, $i = 1, ..., K$, and write the problem as

$$\sum_{j=1}^{J} A_{ij}x_j + x_{J+i} = b_i, \tag{11.10}$$

for $i = 1, ..., K$,

$$\sum_{j=1}^{J} A_{ij}x_j = b_i, \tag{11.11}$$

for $i = K + 1, ..., I$,

$$\sum_{j=1}^{J} Q_{mj}x_j + \sum_{i=1}^{I} A_{im}\mu_i - \nu_m = -c_m, \tag{11.12}$$

for $m = 1, ..., J$,

$$\mu_i x_{J+i} = 0,$$

for $i = 1, ..., K$, and

$$x_j \nu_j = 0,$$

for $j = 1, ..., J$. The objective now is to formulate the problem as a primal linear-programming problem in standard form.

The variables x_j and ν_j, for $j = 1, ..., J$, and μ_i and x_{J+i}, for $i = 1, ..., K$, must be nonnegative; the variables μ_i are unrestricted, for $i = K+1, ..., I$, so for these variables we write

$$\mu_i = \mu_i^+ - \mu_i^-,$$

and require that both μ_i^+ and μ_i^- be nonnegative. Finally, we need a linear functional to minimize.

We rewrite Equation (11.10) as

$$\sum_{j=1}^{J} A_{ij} x_j + x_{J+i} + y_i = b_i, \tag{11.13}$$

for $i = 1, ..., K$, Equation (11.11) as

$$\sum_{j=1}^{J} A_{ij} x_j + y_i = b_i, \tag{11.14}$$

for $i = K+1, ..., I$, and Equation (11.12) as

$$\sum_{j=1}^{J} Q_{mj} x_j + \sum_{i=1}^{I} A_{im} \mu_i - \nu_m + y_{I+m} = -c_m, \tag{11.15}$$

for $m = 1, ..., J$. In order for all the equations to hold, each of the y_i must be zero. The linear programming problem is therefore to minimize the linear functional

$$y_1 + ... + y_{I+J},$$

over nonnegative y_i, subject to the equality constraints in the equations (11.13), (11.14), and (11.15). Any solution to the original problem must be a basic feasible solution to this primal linear-programming problem. Wolfe's Algorithm [176] is a modification of the simplex method that guarantees the complementary slackness conditions; that is, we never have μ_i and x_{J+i} positive basic variables at the same time, nor x_j and ν_j.

11.14.2 An Example

The following example is taken from [176]. Minimize the function

$$f(x_1, x_2) = x_1^2 - x_1 x_2 + 2x_2^2 - x_1 - x_2,$$

subject to the constraints

$$x_1 - x_2 \geq 3,$$

and

$$x_1 + x_2 = 4.$$

We introduce the slack variable x_3 and then minimize

$$y_1 + y_2 + y_3 + y_4,$$

subject to $y_i \geq 0$, for $i = 1, ..., 4$, and the equality constraints

$$x_1 - x_2 - x_3 + y_1 = 3,$$

$$x_1 + x_2 + y_2 = 4,$$

$$2x_1 - x_2 - \mu_1 + \mu_2^+ - \mu_2^- - \nu_1 + y_3 = 1,$$

and

$$-x_1 + 4x_2 + \mu_1 + \mu_2^+ - \mu_2^- - \nu_2 + y_4 = 1.$$

This problem is then solved using the simplex algorithm, modified according to Wolfe's Algorithm.

11.14.3 Equality Constraints

We turn now to the particular case of QP in which all the constraints are equations. The problem is, therefore, to minimize

$$f(x) = a + x^T c + \frac{1}{2} x^T Q x,$$

subject to the constraints

$$Ax = b.$$

The KKT Theorem then tells us that there is λ^* so that $\nabla L(x^*, \lambda^*) = 0$ for the solution vector x^*. Therefore, we have

$$Qx^* + A^T \lambda^* = -c,$$

and

$$Ax^* = b.$$

Such quadratic programming problems arise in sequential quadratic programming.

11.14.4 Sequential Quadratic Programming

Consider once again the CP problem of minimizing the convex function $f(x)$, subject to $g_i(x) = 0$, for $i = 1, ..., I$. The Lagrangian is

$$L(x, \lambda) = f(x) + \sum_{i=1}^{I} \lambda_i g_i(x).$$

We assume that a sensitivity vector λ^* exists, so that x^* solves our problem if and only if (x^*, λ^*) satisfies

$$\nabla L(x^*, \lambda^*) = 0.$$

The problem can then be formulated as finding a zero of the function

$$G(x, \lambda) = \nabla L(x, \lambda),$$

and the Newton–Raphson iterative algorithm can be applied. Because we are modifying both x and λ, this is a primal-dual algorithm.

One step of the Newton–Raphson algorithm has the form

$$\begin{bmatrix} x^{k+1} \\ \lambda^{k+1} \end{bmatrix} = \begin{bmatrix} x^k \\ \lambda^k \end{bmatrix} + \begin{bmatrix} p^k \\ v^k \end{bmatrix},$$

where

$$\begin{bmatrix} \nabla^2_{xx} L(x^k, \lambda^k) & \nabla g(x^k) \\ \nabla g(x^k)^T & 0 \end{bmatrix} \begin{bmatrix} p^k \\ v^k \end{bmatrix} = \begin{bmatrix} -\nabla_x L(x^k, \lambda^k) \\ -g(x^k) \end{bmatrix}.$$

The incremental vector $\begin{bmatrix} p^k \\ v^k \end{bmatrix}$ obtained by solving this system is also the solution to the quadratic-programming problem of minimizing the function

$$\frac{1}{2} p^T \nabla^2_{xx} L(x^k, \lambda^k) p + p^T \nabla_x L(x^k, \lambda^k),$$

subject to the constraint

$$\nabla g(x^k)^T p + g(x^k) = 0.$$

Therefore, the Newton–Raphson algorithm for the original minimization problem can be implemented as a sequence of quadratic programs, each solved by the methods discussed previously. In practice, variants of this approach that employ approximations for the first and second partial derivatives are often used.

11.15 Simulated Annealing

In this chapter we have focused on the minimization of convex functions. For such functions, a local minimum is necessarily a global one. For nonconvex functions, this is not the case. For example, the function $f(x) = x^4 - 8x^3 + 20x^2 - 16.5x + 7$ has a local minimum around $x = 0.6$ and a global minimum around $x = 3.5$. The descent methods we have discussed can get caught at a local minimum that is not global, since we insist on always taking a step that reduces $f(x)$. The *simulated annealing algorithm* [1, 158], also called the *Metropolis algorithm*, is sometimes able to avoid being trapped at a local minimum by permitting an occasional step that increases $f(x)$. The name comes from the analogy with the physical problem of lowering the energy of a solid by first raising the temperature, to bring the particles into a disorganized state, and then gradually reducing the temperature, so that a more organized state is achieved.

Suppose we have calculated x^k. We now generate a random direction and a small random step length. If the new vector $x^k + \Delta x$ makes $f(x)$ smaller, we accept the vector as x^{k+1}. If not, then we accept this vector, with probability

$$Prob(\text{accept}) = \exp\Big(\frac{f(x^k) - f(x^k + \Delta x)}{c_k}\Big),$$

where $c_k > 0$, known as the *temperature*, is chosen by the user. As the iteration proceeds, the temperature c_k is gradually reduced, making it easier to accept increases in $f(x)$ early in the process, but harder later. How to select the temperatures is an art, not a science.

11.16 Exercises

Ex. 11.1 *Prove Lemma 11.1.*

Ex. 11.2 *Apply the Newton–Raphson method to obtain an iterative procedure for finding $\sqrt{a}$, for any positive a. For which x^0 does the method converge? There are two answers, of course; how does the choice of x^0 determine which square root becomes the limit?*

Ex. 11.3 *Apply the Newton–Raphson method to obtain an iterative procedure for finding $a^{1/3}$, for any real a. For which x^0 does the method converge?*

Ex. 11.4 *Extend the Newton–Raphson method to complex variables. Redo the previous exercises for the case of complex a. For the complex case, a has two square roots and three cube roots. How does the choice of x^0 affect the limit? Warning: The case of the cube root is not as simple as it may appear, and has a close connection to fractals and chaos; see [186].*

Ex. 11.5 *Use the reduced Newton–Raphson method to minimize the function $\frac{1}{2}x^TQx$, subject to $Ax = b$, where*

$$Q = \begin{bmatrix} 0 & -13 & -6 & -3 \\ -13 & 23 & -9 & 3 \\ -6 & -9 & -12 & 1 \\ -3 & 3 & 1 & -1 \end{bmatrix},$$

$$A = \begin{bmatrix} 2 & 1 & 2 & 1 \\ 1 & 1 & 3 & -1 \end{bmatrix},$$

and

$$b = \begin{bmatrix} 3 \\ 2 \end{bmatrix}.$$

Start with

$$x^0 = \begin{bmatrix} 1 \\ 1 \\ 0 \\ 0 \end{bmatrix}.$$

Ex. 11.6 *Use the reduced steepest descent method with an exact line search to solve the problem in the previous exercise.*

Chapter 12

Solving Systems of Linear Equations

12.1 Chapter Summary

Optimization plays an important role in solving systems of linear equations. In many applications the linear system is under-determined, meaning that there are multiple, indeed, infinitely many, solutions to the system. It is natural, then, to seek a solution that is optimal, in some sense. When the system involves measured data, as is often the case, there may be no exact solution, or an exact solution to the system may be too noisy. Then an approximate solution, or a solution to a related *regularized* system is sought. In this chapter we discuss briefly both of these situations, focusing on iterative algorithms that have been designed for such problems. For a more in-depth analysis of these problems see [59].

12.2 Arbitrary Systems of Linear Equations

We begin by considering systems of the form $Ax = b$, where A is a real M by N matrix, b a real M by 1 vector, and x is the N by 1 solution vector being sought. If the system has solutions, if there are no additional constraints being imposed on x, and if M and N are not too large, standard noniterative methods, such as Gauss elimination, can be used to find a solution. In all other cases, iterative methods are usually needed.

12.2.1 Under-Determined Systems of Linear Equations

Suppose that $Ax = b$ is a consistent linear system of M equations in N unknowns, where $M < N$. Then there are infinitely many solutions. A standard procedure in such cases is to find that solution x having the smallest two-norm

$$||x||_2 = \sqrt{\sum_{n=1}^{N} |x_n|^2}.$$

As we shall see shortly, the *minimum two-norm* solution of $Ax = b$ is a vector of the form $x = A^T z$, where A^T denotes the transpose of the matrix A. Then $Ax = b$ becomes $AA^T z = b$. Typically, $(AA^T)^{-1}$ will exist, and we get $z = (AA^T)^{-1}b$, from which it follows that the minimum norm solution is $x = A^T(AA^T)^{-1}b$. When M and N are not too large, forming the matrix AA^T and solving for z is not prohibitively expensive and time-consuming. However, in image processing the vector x is often a vectorization of a two-dimensional (or even three-dimensional) image and M and N can be on the order of tens of thousands or more. The ART algorithm gives us a fast method for finding the minimum norm solution without computing AA^T.

We begin by describing the minimum two-norm solution of a consistent system $Ax = b$.

Definition 12.1 *The* null space *of an M by N matrix A is the subspace of $\mathbb{R}^N$ consisting of all w such that $Aw = 0$.*

Theorem 12.1 *The minimum two-norm solution of $Ax = b$ has the form $x = A^T z$ for some z in $\mathbb{R}^M$.*

Proof: If $Ax = b$ then $A(x + w) = b$ for all w in the null space of A. If $x = A^T z$ and w is in the null space of A, then

$$||x + w||_2^2 = ||A^T z + w||_2^2 = (A^T z + w)^T (A^T z + w)$$

$$= (A^T z)^T (A^T z) + (A^T z)^T w + w^T (A^T z) + w^T w$$

$$= ||A^T z||_2^2 + (A^T z)^T w + w^T (A^T z) + ||w||_2^2$$

$$= ||A^T z||_2^2 + ||w||_2^2,$$

since

$$w^T (A^T z) = (Aw)^T z = 0^T z = 0$$

and

$$(A^T z)^T w = z^T Aw = z^T 0 = 0.$$

Therefore, $||x + w||_2 = ||A^T z + w||_2 > ||A^T z||_2 = ||x||_2$ unless $w = 0$. This completes the proof. ∎

12.2.2 Over-Determined Systems of Linear Equations

When the system $Ax = b$ has no solutions, we can look for approximate solutions. For example, we can calculate a vector x for which the function

$$f(x) = \frac{1}{2}||Ax - b||_2^2$$

is minimized; such a vector is called a *least-squares* solution. Setting the gradient equal to zero, we obtain

$$0 = \nabla f(x) = A^T (Ax - b),$$

so that

$$x = (A^T A)^{-1} A^T b,$$

provided that $A^T A$ is invertible, which is usually the case.

12.2.3 Landweber's Method

Landweber's iterative method [144] has the following iterative step: For $k = 0, 1, ...$ let

$$x^{k+1} = x^k + \gamma A^T(b - Ax^k),$$

where A^T denotes the transpose of the matrix A. If the parameter γ is chosen to lie within the interval $(0, 2/L)$, where L is the largest eigenvalue of the matrix A^TA, then the sequence $\{x^k\}$ converges to the solution of $Ax = b$ for which $\|x - x^0\|_2$ is minimized, provided that solutions exist. If not, the sequence $\{x^k\}$ converges to a *least-squares* solution: The limit is the minimizer of the function $\|b - Ax\|_2$ for which $\|x - x^0\|_2$ is minimized.

A least-squares solution of $Ax = b$ is an exact solution of the system

$$A^TAx = A^Tb.$$

One advantage to using Landweber's algorithm is that we do not have to use the matrix A^TA, which can be time-consuming to calculate when M and N are large. As discussed in [59], reasonable estimates of L can also be obtained without knowing A^TA.

12.2.4 The Projected Landweber Algorithm

Suppose that C is a nonempty, closed and convex subset of $\mathbb{R}^N$, and we want to find an exact or approximate solution of $Ax = b$ within C. The *projected Landweber algorithm* (PLW) has the following iterative step:

$$x^{k+1} = P_C\Big(x^k + \gamma A^T(b - Ax^k)\Big),$$

where P_Cx denotes the orthogonal projection of x onto C.

Theorem 12.2 *If the parameter γ is chosen to lie within the interval $(0, 2/L)$, the sequence $\{x^k\}$ converges to an x in C that solves $Ax = b$, provided that solutions exist in C. If not, the sequence $\{x^k\}$ converges to a minimizer, over x in C, of the function $\|b - Ax\|$, if such a minimizer exists.*

Proof: Suppose that $z \in C$ minimizes $f(x) = \frac{1}{2}\|b - Ax\|^2$, over all $x \in C$. Then we have

$$z = P_C(z - \gamma A^T(Az - b)).$$

Therefore,

$$\|z - x^{k+1}\|^2 = \|P_C(z - \gamma A^T(Az - b)) - P_C(x^k - \gamma A^T(Ax^k - b))\|^2$$

$$\leq \|(z - \gamma A^T(Az - b)) - (x^k - \gamma A^T(Ax^k - b))\|^2 = \|z - x^k + \gamma A^T(Ax^k - Az)\|^2$$

$$= \|z - x^k\|^2 + 2\gamma\langle z - x^k, A^T(Ax^k - Az)\rangle + \gamma^2\|A^T(Ax^k - Az)\|^2$$
$$\leq \|z - x^k\|^2 - 2\gamma\|Az - Ax^k\|^2 + \gamma^2\|A^T\|^2\|Az - Ax^k\|^2$$
$$= \|z - x^k\|^2 - (2\gamma - \gamma^2 L)\|Az - Ax^k\|^2.$$

So we have

$$\|z - x^k\|^2 - \|z - x^{k+1}\|^2 \geq (2\gamma - \gamma^2 L)\|Az - Ax^k\|^2 \geq 0.$$

Consequently, we have that the sequence $\{\|z - x^k\|\}$ is decreasing, the sequence $\{\|Az - Ax^k\|\}$ converges to zero, the sequence $\{x^k\}$ is bounded, and a subsequence converges to some $x^* \in C$, with $Ax^* = Az$. It follows that $\{\|x^* - x^k\|\}$ converges to zero, so that $\{x^k\}$ converges to x^*, which is a minimizer of $f(x)$ over $x \in C$. ∎

12.2.5 The Split-Feasibility Problem

Suppose now that C and Q are nonempty, closed and convex subsets of $\mathbb{R}^N$ and $\mathbb{R}^M$, respectively, and we want x in C for which Ax is in Q; this is the *split-feasibility problem* (SFP) [71]. The CQ algorithm [50, 51] has the following iterative step:

$$x^{k+1} = P_C\Big(x^k - \gamma A^T(I - P_Q)Ax^k\Big).$$

For γ in the interval $(0, 2/L)$, the sequence $\{x^k\}$ generated by the CQ algorithm converges to a solution of the SFP, when solutions exist. If no solution exists, it converges to a minimizer, over x in C, of the function

$$f(x) = \frac{1}{2}\|P_Q Ax - Ax\|_2^2, \tag{12.1}$$

provided such minimizers exist. Both the Landweber and projected Landweber methods are special cases of the CQ algorithm.

The following theorem describes the gradient of the function $f(x)$ in Equation (12.1).

Theorem 12.3 *Let $f(x) = \frac{1}{2}\|P_Q Ax - Ax\|_2^2$ and $t \in \partial f(x)$. Then $t = A^T(I - P_Q)Ax$, so that $t = \nabla f(x)$.*

Proof: First, we show that $t = A^T z^*$ for some z^*. Let $s = x + w$, where w is an arbitrary member of the null space of A. Then $As = Ax$ and $f(s) = f(x)$. From

$$0 = f(s) - f(x) \geq \langle t, s - x\rangle = \langle t, w\rangle,$$

it follows that

$$\langle t, w\rangle = 0,$$

for all w in the null space of A, from which we conclude that t is in the range of A^T. Therefore, we can write $t = A^T z^*$.

Let u be chosen so that $\|A(u-x)\| = 1$, and let $\epsilon > 0$. We then have

$$\|P_Q Ax - A(x + \epsilon(u-x))\|^2 - \|P_Q Ax - Ax\|^2 \geq$$

$$\|P_Q(Ax + \epsilon(u-x)) - A(x + \epsilon(u-x))\|^2 - \|P_Q Ax - Ax\|^2 \geq 2\epsilon\langle t, u-x\rangle.$$

Therefore, since

$$\|P_Q Ax - A(x+\epsilon(u-x))\|^2 = \|P_Q Ax - Ax\|^2 - 2\epsilon\langle P_Q Ax - Ax, A(u-x)\rangle + \epsilon^2,$$

it follows that

$$\frac{\epsilon}{2} \geq \langle P_Q Ax - Ax + z^*, A(u-x)\rangle = -\langle A^T(I - P_Q)Ax - t, u-x\rangle.$$

Since ϵ is arbitrary, it follows that

$$\langle A^T(I - P_Q)Ax - t, u - x\rangle \geq 0,$$

for all appropriate u. But this is also true if we replace u with $v = 2x - u$. Consequently, we have

$$\langle A^T(I - P_Q)Ax - t, u - x\rangle = 0.$$

Now we select

$$u - x = (A^T(I - P_Q)Ax - t)/\|AA^T(I - P_Q)Ax - At\|,$$

from which it follows that

$$A^T(I - P_Q)Ax = t.$$

∎

Corollary 12.1 *The gradient of the function*

$$f(x) = \frac{1}{2}\|x - P_C x\|^2$$

is $\nabla f(x) = x - P_C x$, *and the gradient of the function*

$$g(x) = \frac{1}{2}\Big(\|x\|_2^2 - \|x - P_C x\|_2^2\Big)$$

is $\nabla g(x) = P_C x$.

Extensions of the CQ algorithm have been applied recently to problems in intensity-modulated radiation therapy [69, 73].

12.2.6 An Extension of the CQ Algorithm

Let $C \in \mathbb{R}^N$ and $Q \in \mathbb{R}^M$ be closed, nonempty convex sets, and let A and B be J by N and J by M real matrices, respectively. The problem is to find $x \in C$ and $y \in Q$ such that $Ax = By$. When there are no such x and y, we consider the problem of minimizing

$$f(x, y) = \frac{1}{2}\|Ax - By\|_2^2,$$

over $x \in C$ and $y \in Q$.

Let $K = C \times Q$ in $\mathbb{R}^N \times \mathbb{R}^M$. Define

$$G = \begin{bmatrix} A & -B \end{bmatrix},$$

$$w = \begin{bmatrix} x \\ y \end{bmatrix},$$

so that

$$G^T G = \begin{bmatrix} A^T A & -A^T B \\ -B^T A & B^T B \end{bmatrix}.$$

The original problem can now be reformulated as finding $w \in K$ with $Gw = 0$. We shall consider the more general problem of minimizing the function $\|Gw\|$ over $w \in K$. The projected Landweber algorithm (PLW) solves this more general problem.

The iterative step of the PLW algorithm is the following:

$$w^{k+1} = P_K(w^k - \gamma G^T(Gw^k)).$$

Expressing this in terms of x and y, we obtain

$$x^{k+1} = P_C(x^k - \gamma A^T(Ax^k - By^k));$$

and

$$y^{k+1} = P_Q(y^k + \gamma B^T(Ax^k - By^k)).$$

The PLW converges, in this case, to a minimizer of $\|Gw\|$ over $w \in K$, whenever such minimizers exist, for $0 < \gamma < \frac{2}{\rho(G^T G)}$.

12.2.7 The Algebraic Reconstruction Technique

The algorithms presented previously in this chapter are *simultaneous* methods, meaning that all the equations of the system are used at each step of the iteration. Such methods tend to converge slowly, which presents a major problem for large systems. The *algebraic reconstruction technique* (ART) is a *row-action* method, meaning that only a single equation is used at each step of the iteration. The ART has the following iterative step: For $k = 0, 1, ...$ and $m = k(\mathrm{mod}\, M) + 1$, let

$$x_n^{k+1} = x_n^k + A_{mn}(b_m - (Ax^k)_m)/\sum_{j=1}^{N} |A_{mj}|^2.$$

We can describe the ART geometrically as follows: Once we have x^k and m, the vector x^{k+1} is the orthogonal projection of x^k onto the hyperplane H_m given by

$$H_m = \{x|(Ax)_m = b_m\}.$$

The Landweber algorithm can be similarly described: The vector x^{k+1} is a weighted sum of the orthogonal projections of x^k onto each of the hyperplanes H_m, for all m.

In the consistent case, when the system $Ax = b$ has solutions, the ART converges to the solution for which $\|x - x^0\|$ is minimized. Unlike the simultaneous methods, when no solution exists, the ART sequence $\{x^k\}$ does not converge to a single vector, but subsequences do converge to members of a *limit cycle* consisting of (typically) M distinct vectors. Generally speaking, the ART will converge, in the consistent case, faster than the Landweber method, especially if the equations are selected in a random order [127].

12.2.8 Double ART

Because the ART is significantly faster to converge than the Landweber method in the consistent case, we would like to be able to use the ART in the inconsistent case, as well, to get a least-squares solution. To avoid the limit-cycle behavior of ART in this case, we can use *double ART* (DART).

We know from basic linear algebra that the vector b can be written as

$$b = A\hat{x} + \hat{w},$$

where $\hat{x}$ minimizes the function $\|b - Ax\|_2$ and $w = \hat{w}$ minimizes the function $\|b - w\|_2$, subject to $A^T w = 0$. Said another way, $A\hat{x}$ is the orthogonal projection of b onto the range of A and $\hat{w}$ is the orthogonal projection of b onto the null space of A^T.

In DART we apply the ART algorithm twice, first to the consistent linear system $A^T w = 0$, with $w^0 = b$, so that the limit is $\hat{w}$, and then to the consistent system $Ax = b - \hat{w}$. The result is the minimizer of $\|b - Ax\|$ for which $\|x - x^0\|$ is minimized.

12.3 Regularization

In many applications in which systems of linear equations must be solved, the entries of the vector b are measured data and $Ax = b$ is a model that attempts to describe, in a somewhat simplified way, how b depends on the unknown vector x. The statistical noise in the measured data introduces one type of error, while the approximate nature of the model itself introduces another. Because the model is simplified, but the data b is noisy, an exact solution x itself usually ends up noisy. Also, it is common for the system to be *ill-conditioned*, that is, for small changes in b to lead to large changes in the exact solution x. This happens when the ratio of the largest to smallest eigenvalues of the matrix A^TA is large. In such cases even a minimum-norm solution of $Ax = b$ can have a large norm. Consequently, we often do not want an exact solution of $Ax = b$, even when such solutions exist. Instead, we *regularize* the problem.

12.3.1 Norm-Constrained Least-Squares

One way to regularize the problem is to minimize not $\|b - Ax\|_2$, but, say,

$$f(x) = \|b - Ax\|_2^2 + \epsilon^2\|x\|_2^2, \tag{12.2}$$

for some small $\epsilon > 0$. Now we are still trying to make $\|b - Ax\|_2$ small, but managing to keep $\|x\|_2$ from becoming too large in the process. This leads to a *norm-constrained least-squares* solution.

The minimizer of $f(x)$ is the unique solution $\hat{x}_\epsilon$ of the system

$$(A^TA + \epsilon^2 I)x = A^Tb.$$

When M and N are large, we need ways to solve this system without having to deal with the matrix $A^TA + \epsilon^2 I$. Landweber's method allowed us to avoid A^TA in calculating the least-squares solution. Is there a similar method to use now? Yes, there is.

12.3.2 Regularizing Landweber's Algorithm

Our goal is to minimize the function $f(x)$ in Equation (12.2). Notice that this is equivalent to minimizing the function

$$F(x) = ||Bx - c||_2^2,$$

for

$$B = \begin{bmatrix} A \\ \epsilon I \end{bmatrix},$$

and

$$c = \begin{bmatrix} b \\ 0 \end{bmatrix},$$

where 0 denotes a column vector with all entries equal to zero. The Landweber iteration for the problem $Bx = c$ is

$$x^{k+1} = x^k + \alpha B^T(c - Bx^k), \tag{12.3}$$

for $0 < \alpha < 2/\rho(B^TB)$, where $\rho(B^TB)$ is the largest eigenvalue, or the spectral radius, of B^TB. Equation (12.3) can be written as

$$x^{k+1} = (1 - \alpha\epsilon^2)x^k + \alpha A^T(b - Ax^k).$$

12.3.3 Regularizing the ART

We would like to get the regularized solution $\hat{x}_\epsilon$ by taking advantage of the faster convergence of the ART. Fortunately, there are ways to find $\hat{x}_\epsilon$, using only the matrix A and the ART algorithm. We discuss two methods for using ART to obtain regularized solutions of $Ax = b$. The first one is presented in [53], while the second one is due to Eggermont, Herman, and Lent [105].

In our first method we use ART to solve the system of equations given in matrix form by

$$\begin{bmatrix} A^T & \epsilon I \end{bmatrix} \begin{bmatrix} u \\ v \end{bmatrix} = 0. \tag{12.4}$$

We begin with $u^0 = b$ and $v^0 = 0$. Then, the lower component of the limit vector is $v^\infty = -\epsilon\hat{x}_\epsilon$, while the upper limit is $u^\infty = b - A\hat{x}_\epsilon$.

The method of Eggermont *et al.* is similar. In their method we use ART to solve the system of equations given in matrix form by

$$\begin{bmatrix} A & \epsilon I \end{bmatrix} \begin{bmatrix} x \\ v \end{bmatrix} = b. \tag{12.5}$$

We begin at $x^0 = 0$ and $v^0 = 0$. Then, the limit vector has for its upper component $x^\infty = \hat{x}_\epsilon$, and $\epsilon v^\infty = b - A\hat{x}_\epsilon$.

12.4 Nonnegative Systems of Linear Equations

We turn now to nonnegative systems of linear equations, which we shall denote by $y = Px$, with the understanding that P is an I by J matrix with nonnegative entries P_{ij}, such that, for each j, the column sum

$$s_j = \sum_{i=1}^{I} P_{ij}$$

is positive, y is an I by 1 vector with positive entries y_i, and we seek a solution x with nonnegative entries x_j. We say that the system is *consistent* whenever such nonnegative solutions exist. Denote by $\mathcal{X}$ the set of all nonnegative x for which the vector Px has only positive entries. In what follows, all vectors x will lie in $\mathcal{X}$ and the initial vector x^0 will always be positive.

12.4.1 The Multiplicative ART

Both the algebraic reconstruction technique (ART) and the *multiplicative algebraic reconstruction technique* (MART) were introduced by Gordon, Bender and Herman [121] as two iterative methods for discrete image reconstruction in transmission tomography. It was noticed somewhat later that the ART is a special case of Kaczmarz's algorithm [134].

Both methods are what are called *row-action* methods, meaning that each step of the iteration uses only a single equation from the system. The MART is limited to nonnegative systems for which nonnegative solutions are sought. In the under-determined case, both algorithms find the solution closest to the starting vector, in the two-norm or weighted two-norm sense for ART, and in the cross-entropy sense for MART, so both algorithms can be viewed as solving optimization problems. We consider two different versions of the MART.

12.4.2 MART I

The iterative step of the first version of MART, which we call MART I, is the following: For $k = 0, 1, ...$, and $i = k(\text{mod}\, I) + 1$, let

$$x_j^{k+1} = x_j^k \Big(\frac{y_i}{(Px^k)_i} \Big)^{P_{ij}/m_i},$$

for $j = 1, ..., J$, where the parameter m_i is defined to be

$$m_i = \max\{P_{ij} | j = 1, ..., J\}.$$

The MART I algorithm converges, in the consistent case, to the nonnegative solution for which the KL distance $KL(x, x^0)$ is minimized.

12.4.3 MART II

The iterative step of the second version of MART, which we shall call MART II, is the following: For $k = 0, 1, ...,$ and $i = k(\mathrm{mod}\, I) + 1$, let

$$x_j^{k+1} = x_j^k \Big(\frac{y_i}{(Px^k)_i}\Big)^{P_{ij}/s_j n_i},$$

for $j = 1, ..., J$, where the parameter n_i is defined to be

$$n_i = \max\{P_{ij} s_j^{-1} | j = 1, ..., J\}.$$

The MART II algorithm converges, in the consistent case, to the nonnegative solution for which the KL distance

$$\sum_{j=1}^{J} s_j KL(x_j, x_j^0)$$

is minimized. Just as Landweber's method is a simultaneous cousin of the row-action ART, there is a simultaneous cousin of the MART, called, not surprisingly, the *simultaneous MART* (SMART).

12.4.4 The Simultaneous MART

The SMART minimizes the cross-entropy, or Kullback–Leibler distance, $f(x) = KL(Px, y)$, over nonnegative vectors x [93, 81, 185, 39].

Having found the vector x^k, the next vector in the SMART sequence is x^{k+1}, with entries given by

$$x_j^{k+1} = x_j^k \exp\Bigg(s_j^{-1} \sum_{i=1}^{I} P_{ij} \log\Big(\frac{y_i}{(Px^k)_i}\Big)\Bigg).$$

As with MART II, when there are nonnegative solutions of $y = Px$, the SMART converges to the solution for which the KL distance

$$\sum_{j=1}^{J} s_j KL(x_j, x_j^0)$$

is minimized.

12.4.5 The EMML Iteration

The *expectation maximization maximum likelihood* algorithm (EMML) minimizes the function $f(x) = KL(y, Px)$, over nonnegative vectors x [187,

145, 200, 146, 39]. Having found the vector x^k, the next vector in the EMML sequence is x^{k+1}, with entries given by

$$x_j^{k+1} = x_j^k s_j^{-1} \left(\sum_{i=1}^{I} P_{ij} \left(\frac{y_i}{(Px^k)_i} \right) \right).$$

The iterative step of the EMML is closely related to that of the SMART, except that the exponentiation and logarithm are missing. When there are nonnegative solutions of the system $y = Px$, the EMML converges to a nonnegative solution, but no further information about this solution is known. Both the SMART and the EMML are slow to converge, particularly when the system is large.

12.4.6 Alternating Minimization

In [39] the SMART and the EMML were derived using the following *alternating minimization* approach.

For each $x \in \mathcal{X}$, let $r(x)$ and $q(x)$ be the I by J arrays with entries

$$r(x)_{ij} = x_j P_{ij} y_i / (Px)_i,$$

and

$$q(x)_{ij} = x_j P_{ij}.$$

In the iterative step of the SMART we get x^{k+1} by minimizing the function

$$KL(q(x), r(x^k)) = \sum_{i=1}^{I} \sum_{j=1}^{J} KL(q(x)_{ij}, r(x^k)_{ij})$$

over $x \geq 0$. Note that $KL(Px, y) = KL(q(x), r(x))$. Similarly, the iterative step of the EMML is to minimize the function $KL(r(x^k), q(x))$ to get $x = x^{k+1}$. Note that $KL(y, Px) = KL(r(x), q(x))$.

12.4.7 The Row-Action Variant of EMML

When there are nonnegative solutions of $y = Px$, the MART converges faster than the SMART, and to the same solution. The SMART involves exponentiation and a logarithm, and the MART a non-integral power, both of which complicate their calculation. The EMML is considerably simpler in this respect, but, like SMART, converges slowly. We would like to have a row-action variant of the EMML that converges faster than the EMML in the consistent case, but is easier to calculate than the MART. The EMART is such an algorithm. As with the MART, we distinguish two versions,

EMART I and EMART II. When the system $y = Px$ has nonnegative solutions, both EMART I and EMART II converge to nonnegative solutions, but nothing further is known about these solutions. To motivate these algorithms, we rewrite the MART algorithms.

The iterative step of MART I can be written as follows: For $k = 0, 1, ...,$ and $i = k(\bmod I) + 1$, let

$$x_j^{k+1} = x_j^k \exp\left(\left(\frac{P_{ij}}{m_i}\right) \log\left(\frac{y_i}{(Px^k)_i}\right)\right),$$

or, equivalently, as

$$\log x_j^{k+1} = \left(1 - \frac{P_{ij}}{m_i}\right) \log x_j^k + \left(\frac{P_{ij}}{m_i}\right) \log\left(x_j^k \frac{y_i}{(Px^k)_i}\right). \tag{12.6}$$

Similarly, the iterative step of MART II can be written as follows: For $k = 0, 1, ...,$ and $i = k(\bmod I) + 1$, let

$$x_j^{k+1} = x_j^k \exp\left(\left(\frac{P_{ij}}{s_j n_i}\right) \log\left(\frac{y_i}{(Px^k)_i}\right)\right),$$

or, equivalently, as

$$\log x_j^{k+1} = \left(1 - \frac{P_{ij}}{s_j n_i}\right) \log x_j^k + \left(\frac{P_{ij}}{s_j n_i}\right) \log\left(x_j^k \frac{y_i}{(Px^k)_i}\right). \tag{12.7}$$

We obtain the EMART I and EMART II simply by removing the logarithms in Equations (12.6) and (12.7), respectively.

12.4.8 EMART I

The iterative step of EMART I is as follows: For $k = 0, 1, ...,$ and $i = k(\bmod I) + 1$, let

$$x_j^{k+1} = \left(1 - \frac{P_{ij}}{m_i}\right) x_j^k + \left(\frac{P_{ij}}{m_i}\right)\left(x_j^k \frac{y_i}{(Px^k)_i}\right).$$

12.4.9 EMART II

The iterative step of EMART II is as follows:

$$x_j^{k+1} = \left(1 - \frac{P_{ij}}{s_j n_i}\right) x_j^k + \left(\frac{P_{ij}}{s_j n_i}\right)\left(x_j^k \frac{y_i}{(Px^k)_i}\right).$$

12.5 Regularized SMART and EMML

As with the Landweber algorithm, there are situations that arise in practice in which, because of noisy measurements, the exact or approximate solutions of $y = Px$ provided by the SMART and EMML algorithms are not suitable. In such cases, we need to regularize the SMART and the EMML, which is usually done by including a penalty function.

12.5.1 Regularized SMART

As we have seen, the iterative step of the SMART is obtained by minimizing the function $KL(q(x), r(x^k))$ over nonnegative x, and the limit of the SMART minimizes $KL(Px, y)$. We can regularize by minimizing

$$KL(Px, y) + KL(x, p),$$

where the vector p with positive entries p_j is a prior estimate of the solution. To obtain x^{k+1} from x^k, we minimize

$$KL(q(x), r(x^k)) + \sum_{j=1}^{J} \delta_j KL(x_j, p_j).$$

There are many penalty functions we could use here, but the one we have chosen permits the minimizing x^{k+1} to be obtained in closed form.

The iterative step of the regularized SMART is as follows:

$$\log x_j^{k+1} = \frac{\delta_j}{\delta_j + s_j} \log p_j + \frac{1}{\delta_j + s_j} x_j^k \sum_{i=1}^{I} P_{ij} \log\left(\frac{y_i}{(Px^k)_i}\right).$$

12.5.2 Regularized EMML

As we have seen, the iterative step of the EMML is obtained by minimizing the function $KL(r(x^k), q(x))$ over nonnegative x, and the limit of the EMML minimizes $KL(y, Px)$. We can regularize by minimizing

$$KL(y, Px) + KL(p, x).$$

To obtain x^{k+1} from x^k, we minimize

$$KL(r(x^k), q(x)) + \sum_{j=1}^{J} \delta_j KL(p_j, x_j).$$

Again, there are many penalty functions we could use here, but the one we have chosen permits the minimizing x^{k+1} to be obtained in closed form.

The iterative step of the regularized EMML is as follows:

$$x_j^{k+1} = \frac{\delta_j}{\delta_j + s_j} p_j + \frac{1}{\delta_j + s_j} x_j^k \sum_{i=1}^{I} P_{ij} \Big(\frac{y_i}{(Px^k)_i} \Big).$$

12.6 Block-Iterative Methods

The algorithms we have considered in this chapter are either simultaneous algorithms or row-action ones. There are also *block-iterative* variants of MART and ART, in which some, but not all, equations of the system are used at each step. The subsets of equations used at a single step are called *blocks*. Generally speaking, the smaller the blocks, the faster the convergence, in the consistent case. On the other hand, it may be inconvenient, given the architecture of the computer, to deal with only a single equation at each step. By using blocks, we can achieve a compromise between speed of convergence and compatibility with the architecture of the computer. These block-iterative methods are discussed in detail in [59].

12.7 Exercises

Ex. 12.1 *Show that the two algorithms associated with Equations (12.4) and (12.5), respectively, do actually perform as claimed.*

Chapter 13

Conjugate-Direction Methods

13.1 Chapter Summary

Finding the least-squares solution of a possibly inconsistent system of linear equations $Ax = b$ is equivalent to minimizing the quadratic function $f(x) = \frac{1}{2}||Ax - b||_2^2$ and so can be viewed within the framework of optimization. Iterative optimization methods can then be used to provide, or at least suggest, algorithms for obtaining the least-squares solution. The *conjugate gradient method* is one such method. Proofs for the lemmas in this chapter are exercises for the reader.

13.2 Iterative Minimization

Iterative methods for minimizing a real-valued function $f(x)$ over the vector variable x usually take the following form: Having obtained x^{k-1}, a new direction vector d^k is selected, an appropriate scalar $\alpha_k > 0$ is determined and the next member of the iterative sequence is given by

$$x^k = x^{k-1} + \alpha_k d^k. \tag{13.1}$$

Ideally, one would choose the α_k to be the value of α for which the function $f(x^{k-1} + \alpha d^k)$ is minimized. It is assumed that the direction d^k is a *descent direction*; that is, for small positive α the function $f(x^{k-1} + \alpha d^k)$ is strictly decreasing. Finding the optimal value of α at each step of the iteration is difficult, if not impossible, in most cases, and approximate methods, using line searches, are commonly used.

Lemma 13.1 *When x^k is constructed using the optimal α, we have*

$$\nabla f(x^k) \cdot d^k = 0. \tag{13.2}$$

Proof: Differentiate the function $f(x^{k-1} + \alpha d^k)$ with respect to the variable α. ∎

Since the gradient $\nabla f(x^k)$ is orthogonal to the previous direction vector d^k and also because $-\nabla f(x)$ is the direction of greatest decrease of $f(x)$, the choice of $d^{k+1} = -\nabla f(x^k)$ as the next direction vector is a reasonable one. With this choice we obtain Cauchy's *steepest descent method* [151]:

$$x^{k+1} = x^k - \alpha_{k+1} \nabla f(x^k).$$

The steepest descent method need not converge in general and even when it does, it can do so slowly, suggesting that there may be better choices for the direction vectors. For example, the Newton–Raphson method [164] employs the following iteration:

$$x^{k+1} = x^k - \nabla^2 f(x^k)^{-1} \nabla f(x^k),$$

where $\nabla^2 f(x)$ is the Hessian matrix for $f(x)$ at x. To investigate further the issues associated with the selection of the direction vectors, we consider the more tractable special case of quadratic optimization.

13.3 Quadratic Optimization

Let A be an arbitrary real I by J matrix. The linear system of equations $Ax = b$ need not have any solutions, and we may wish to find a least-squares solution $x = \hat{x}$ that minimizes

$$f(x) = \frac{1}{2}||b - Ax||_2^2. \tag{13.3}$$

The vector b can be written

$$b = A\hat{x} + \hat{w},$$

where $A^T\hat{w} = 0$ and a least squares solution is an exact solution of the linear system $Qx = c$, with $Q = A^TA$ and $c = A^Tb$. We shall assume that Q is invertible and there is a unique least squares solution; this is the typical case.

We consider now the iterative scheme described by Equation (13.1) for $f(x)$ as in Equation (13.3). For now, the direction vectors d^k are arbitrary. For this $f(x)$ the gradient becomes

$$\nabla f(x) = Qx - c.$$

The optimal α_k for the iteration can be obtained in closed form.

Lemma 13.2 *The optimal* α_k *is*

$$\alpha_k = \frac{r^k \cdot d^k}{d^k \cdot Qd^k}, \tag{13.4}$$

where $r^k = c - Qx^{k-1}$.

Lemma 13.3 *Let* $||x||_Q^2 = x \cdot Qx$ *denote the square of the Q-norm of* x. *Then*

$$||\hat{x} - x^{k-1}||_Q^2 - ||\hat{x} - x^k||_Q^2 = (r^k \cdot d^k)^2/d^k \cdot Qd^k \geq 0$$

for any direction vectors d^k.

If the sequence of direction vectors $\{d^k\}$ is completely general, the iterative sequence need not converge. However, if the set of direction vectors is finite and spans $\mathbb{R}^J$ and we employ them cyclically, convergence follows.

Theorem 13.1 *Let* $\{d^1, ..., d^J\}$ *be any basis for* $\mathbb{R}^J$. *Let* α_k *be chosen according to Equation (13.4). Then, for* $k = 1, 2, ...,$ $j = k(\text{mod}\, J)$, *and any* x^0, *the sequence defined by*

$$x^k = x^{k-1} + \alpha_k d^j$$

converges to the least squares solution.

Proof: The sequence $\{||\hat{x} - x^k||_Q^2\}$ is decreasing and, therefore, the sequence $\{(r^k \cdot d^k)^2/d^k \cdot Qd^k$ must converge to zero. Therefore, the vectors x^k are bounded, and for each $j = 1, ..., J$, the subsequences $\{x^{mJ+j},\ m = 0, 1, ...\}$ have cluster points, say $x^{*,j}$ with

$$x^{*,j} = x^{*,j-1} + \frac{(c - Qx^{*,j-1}) \cdot d^j}{d^j \cdot Qd^j} d^j.$$

Since

$$r^{mJ+j} \cdot d^j \to 0,$$

it follows that, for each $j = 1, ..., J$,

$$(c - Qx^{*,j}) \cdot d^j = 0.$$

Therefore,

$$x^{*,1} = ... = x^{*,J} = x^*$$

with $Qx^* = c$. Consequently, x^* is the least squares solution and the sequence $\{||x^* - x^k||_Q\}$ is decreasing. But a subsequence converges to zero; therefore, $\{||x^* - x^k||_Q\} \to 0$. This completes the proof. ∎

There is an interesting corollary to this theorem that pertains to a modified version of the ART algorithm. For $k = 0, 1, ...$ and $i = k(\text{mod}\, M) + 1$ and with the rows of A normalized to have length one, the ART iterative step is

$$x^{k+1} = x^k + (b_i - (Ax^k)_i)a^i,$$

where a^i is the ith column of A^T. When $Ax = b$ has no solutions, the ART algorithm does not converge to the least-squares solution; rather, it exhibits subsequential convergence to a limit cycle. However, using the previous theorem, we can show that the following modification of the ART, which we shall call the *least squares ART* (LS-ART), converges to the least-squares solution for every x^0:

$$x^{k+1} = x^k + \frac{r^{k+1} \cdot a^i}{a^i \cdot Qa^i} a^i.$$

In the quadratic case the steepest descent iteration has the form

$$x^k = x^{k-1} + \frac{r^k \cdot r^k}{r^k \cdot Qr^k} r^k.$$

We have the following result.

Theorem 13.2 *The steepest descent method converges to the least-squares solution.*

Proof: As in the proof of the previous theorem, we have

$$||\hat{x} - x^{k-1}||_Q^2 - ||\hat{x} - x^k||_Q^2 = (r^k \cdot d^k)^2 / d^k \cdot Qd^k \geq 0,$$

where now the direction vectors are $d^k = r^k$. So, the sequence $\{||\hat{x} - x^k||_Q^2\}$ is decreasing, and therefore the sequence $\{(r^k \cdot r^k)^2 / r^k \cdot Qr^k\}$ must converge to zero. The sequence $\{x^k\}$ is bounded; let x^* be a cluster point. It follows that $c - Qx^* = 0$, so that x^* is the least-squares solution $\hat{x}$. The rest of the proof follows as in the proof of the previous theorem. ∎

13.4 Conjugate Bases for $\mathbb{R}^J$

If the set $\{v^1, ..., v^J\}$ is a basis for $\mathbb{R}^J$, then any vector x in $\mathbb{R}^J$ can be expressed as a linear combination of the basis vectors; that is, there are real numbers $a_1, ..., a_J$ for which

$$x = a_1 v^1 + a_2 v^2 + ... + a_J v^J.$$

For each x the coefficients a_j are unique. To determine the a_j we write

$$x \cdot v^m = a_1 v^1 \cdot v^m + a_2 v^2 \cdot v^m + ... + a_J v^J \cdot v^m,$$

for $m = 1, ..., J$. Having calculated the quantities $x \cdot v^m$ and $v^j \cdot v^m$, we solve the resulting system of linear equations for the a_j.

If, instead of an arbitrary basis $\{v^1, ..., v^J\}$, we use an orthogonal basis $\{u^1, ..., u^J\}$, that is, then $u^j \cdot u^m = 0$, unless $j = m$, then the system of linear equations is now trivial to solve. The solution is $a_j = x \cdot u^j / u^j \cdot u^j$, for each j. Of course, we still need to compute the quantities $x \cdot u^j$.

The least-squares solution of the linear system of equations $Ax = b$ is

$$\hat{x} = (A^T A)^{-1} A^T b = Q^{-1} c.$$

To express $\hat{x}$ as a linear combination of the members of an orthogonal basis $\{u^1, ..., u^J\}$ we need the quantities $\hat{x} \cdot u^j$, which usually means that we need to know $\hat{x}$ first. For a special kind of basis, a *Q-conjugate basis*, knowing $\hat{x}$ ahead of time is not necessary; we need only know Q and c. Therefore, we can use such a basis to find $\hat{x}$. This is the essence of the *conjugate gradient method* (CGM), in which we calculate a conjugate basis and, in the process, determine $\hat{x}$.

13.4.1 Conjugate Directions

From Equation (13.2) we have

$$(c - Qx^k) \cdot d^k = 0,$$

which can be expressed as

$$(\hat{x} - x^k) \cdot Qd^k = (\hat{x} - x^k)^T Qd^k = 0.$$

Two vectors x and y are said to be *Q-orthogonal* (or *Q-conjugate*, or just *conjugate*), if $x \cdot Qy = 0$. So, the least-squares solution that we seek lies in a direction from x^k that is Q-orthogonal to d^k. This suggests that we can do better than steepest descent if we take the next direction to be Q-orthogonal to the previous one, rather than just orthogonal. This leads us to *conjugate direction methods*.

Definition 13.1 *We say that the set* $\{p^1, ..., p^n\}$ *is a* conjugate set *for* $\mathbb{R}^J$ *if* $p^i \cdot Qp^j = 0$ *for* $i \neq j$.

Lemma 13.4 *A conjugate set that does not contain zero is linearly independent. If* $p^n \neq 0$ *for* $n = 1, ..., J$*, then the least-squares vector* $\hat{x}$ *can be written as*

$$\hat{x} = a_1 p^1 + ... + a_J p^J,$$

with $a_j = c \cdot p^j / p^j \cdot Qp^j$ *for each* j.

Proof: Use the Q-inner product $\langle x, y \rangle_Q = x \cdot Qy$. ∎

Therefore, once we have a conjugate basis, computing the least squares solution is trivial. Generating a conjugate basis can obviously be done using the standard Gram–Schmidt approach.

13.4.2 The Gram–Schmidt Method

Let $\{v^1, ..., v^J\}$ be a linearly independent set of vectors in the space $\mathbb{R}^M$, where $J \leq M$. The Gram–Schmidt method uses the v^j to create an orthogonal basis $\{u^1, ..., u^J\}$ for the span of the v^j. Begin by taking $u^1 = v^1$. For $j = 2, ..., J$, let

$$u^j = v^j - \frac{u^1 \cdot v^j}{u^1 \cdot u^1} u^1 - ... - \frac{u^{j-1} \cdot v^j}{u^{j-1} \cdot u^{j-1}} u^{j-1}.$$

To apply this approach to obtain a conjugate basis, we would simply replace the dot products $u^k \cdot v^j$ and $u^k \cdot u^k$ with the Q-inner products, that is,

$$p^j = v^j - \frac{p^1 \cdot Qv^j}{p^1 \cdot Qp^1} p^1 - ... - \frac{p^{j-1} \cdot Qv^j}{p^{j-1} \cdot Qp^{j-1}} p^{j-1}. \tag{13.5}$$

Even though the Q-inner products can always be written as $x \cdot Qy = Ax \cdot Ay$, so that we need not compute the matrix Q, calculating a conjugate basis using Gram–Schmidt is not practical for large J. There is a way out, fortunately.

If we take $p^1 = v^1$ and $v^j = Qp^{j-1}$, we have a much more efficient mechanism for generating a conjugate basis, namely a three-term recursion formula [151]. The set $\{p^1, Qp^1, ..., Qp^{J-1}\}$ need not be a linearly independent set, in general, but, if our goal is to find $\hat{x}$, and not really to calculate a full conjugate basis, this does not matter, as we shall see.

Theorem 13.3 *Let* $p^1 \neq 0$ *be arbitrary. Let* p^2 *be given by*

$$p^2 = Qp^1 - \frac{Qp^1 \cdot Qp^1}{p^1 \cdot Qp^1} p^1,$$

so that $p^2 \cdot Qp^1 = 0$. Then, for $n \geq 2$, let p^{n+1} be given by

$$p^{n+1} = Qp^n - \frac{Qp^n \cdot Qp^n}{p^n \cdot Qp^n} p^n - \frac{Qp^{n-1} \cdot Qp^n}{p^{n-1} \cdot Qp^{n-1}} p^{n-1}. \tag{13.6}$$

Then, the set $\{p^1, ..., p^J\}$ is a conjugate set for $\mathbb{R}^J$. If $p^n \neq 0$ for each n, then the set is a conjugate basis for $\mathbb{R}^J$.

Proof: We consider the induction step of the proof. Assume that $\{p^1, ..., p^n\}$ is a Q-orthogonal set of vectors; we then show that $\{p^1, ..., p^{n+1}\}$ is also, provided that $n \leq J - 1$. It is clear from Equation (13.6) that

$$p^{n+1} \cdot Qp^n = p^{n+1} \cdot Qp^{n-1} = 0.$$

For $j \leq n - 2$, we have

$$p^{n+1} \cdot Qp^j = p^j \cdot Qp^{n+1} = p^j \cdot Q^2 p^n - ap^j \cdot Qp^n - bp^j \cdot Qp^{n-1},$$

for constants a and b. The second and third terms on the right side are then zero because of the induction hypothesis. The first term is also zero since

$$p^j \cdot Q^2 p^n = (Qp^j) \cdot Qp^n = 0$$

because Qp^j is in the span of $\{p^1, ..., p^{j+1}\}$, and so is Q-orthogonal to p^n. ■

The calculations in the three-term recursion formula Equation (13.6) also occur in the Gram–Schmidt approach in Equation (13.5); the point is that Equation (13.6) uses only the first three terms, in every case.

13.5 The Conjugate Gradient Method

13.5.1 The Main Idea

The main idea in the *conjugate gradient method* (CGM) is to build the conjugate set as we calculate the least squares solution using the iterative algorithm

$$x^n = x^{n-1} + \alpha_n p^n.$$

The α_n is chosen so as to minimize $f(x^{n-1} + \alpha p^n)$, as a function of α. So we have

$$\alpha_n = \frac{r^n \cdot p^n}{p^n \cdot Qp^n},$$

where $r^n = c - Qx^{n-1}$. Since the function $f(x) = \frac{1}{2}||Ax - b||_2^2$ has for its gradient $\nabla f(x) = A^T(Ax - b) = Qx - c$, the residual vector $r^n = c - Qx^{n-1}$ is the direction of steepest descent from the point $x = x^{n-1}$. The CGM combines the use of the negative gradient directions from the steepest descent method with the use of a conjugate basis of directions, by using the r^{n+1} to construct the next direction p^{n+1} in such a way as to form a conjugate set $\{p^1, ..., p^J\}$.

13.5.2 A Recursive Formula

As before, there is an efficient recursive formula that provides the next direction: Let $p^1 = r^1 = (c - Qx^0)$ and for $j = 2, 3, ...$

$$p^j = r^j - \beta_{j-1} p^{j-1}, \tag{13.7}$$

with

$$\beta_{j-1} = \frac{r^j \cdot Qp^{j-1}}{p^{j-1} \cdot Qp^{j-1}}.$$

Note that it follows from the definition of β_{j-1} that

$$p^j Q p^{j-1} = 0. \tag{13.8}$$

Since the α_n is the optimal choice and

$$r^{n+1} = -\nabla f(x^n),$$

we have, according to Equation (13.2),

$$r^{n+1} \cdot p^n = 0.$$

In theory, the CGM converges to the least squares solution in finitely many steps, since we either reach $p^{n+1} = 0$ or $n + 1 = J$. In practice, the CGM can be employed as a fully iterative method by cycling back through the previously used directions.

An induction proof similar to the one used to prove Theorem 13.3 establishes that the set $\{p^1, ..., p^J\}$ is a conjugate set [151, 164]. In fact, we can say more.

Theorem 13.4 *For $n = 1, 2, ..., J$ and $j = 1, ..., n - 1$ we have*

(1) $r^n \cdot r^j = 0$;

(2) $r^n \cdot p^j = 0$; and

(3) $p^n \cdot Qp^j = 0$.

The proof presented here through a series of exercises at the end of the chapter is based on that given in [164].

13.6 Krylov Subspaces

Another approach to deriving the conjugate gradient method is to use Krylov subspaces. If we select $x^0 = 0$ as our starting vector for the CGM, then $p^1 = r^1 = c$, and each p^{n+1} and x^{n+1} lie in the *Krylov subspace* $\mathcal{K}_n(Q, c)$, defined to be the span of the vectors $\{c, Qc, Q^2c, ..., Q^nc\}$.

For any x in $\mathbb{R}^J$, we have

$$\|x - \hat{x}\|_Q^2 = (x - \hat{x})^T Q(x - \hat{x}).$$

Minimizing $\|x - \hat{x}\|_Q^2$ over all x in $\mathcal{K}_n(Q, c)$ is equivalent to minimizing the same function over all x of the form $x = x^n + \alpha p^{n+1}$. This, in turn, is equivalent to minimizing

$$-2\alpha p^{n+1} \cdot r^{n+1} + \alpha^2 p^{n+1} \cdot Qp^{n+1},$$

over all α, which has for its solution the value $\alpha = \alpha_{n+1}$ used to calculate x^{n+1} in the CGM.

13.7 Extensions of the CGM

The convergence rate of the CGM depends on the condition number of the matrix Q, which is the ratio of its largest to its smallest eigenvalues. When the condition number is much greater than one, convergence can be accelerated by *preconditioning* the matrix Q; this means replacing Q with $P^{-1/2}QP^{-1/2}$, for some positive-definite approximation P of Q (see [7]).

There are versions of the CGM for the minimization of non-quadratic functions. In the quadratic case the next conjugate direction p^{n+1} is built from the residual r^{n+1} and p^n. Since, in that case, $r^{n+1} = -\nabla f(x^n)$, this suggests that in the non-quadratic case we build p^{n+1} from $-\nabla f(x^n)$ and p^n. This leads to the Fletcher–Reeves method. Other similar algorithms, such as the Polak–Ribiere and the Hestenes–Stiefel methods, perform better on certain problems [164].

13.8 Exercises

Ex. 13.1 *There are several lemmas in this chapter whose proofs are only sketched. Complete the proofs of these lemma.*

The following exercises refer to the Conjugate Gradient Method.

Ex. 13.2 *Show that*

$$r^{n+1} = r^n - \alpha_n Q p^n, \tag{13.9}$$

so Qp^n is in the span of r^{n+1} and r^n.

Ex. 13.3 *Prove that $r^n = 0$ whenever $p^n = 0$, in which case we have $c = Qx^{n-1}$, so that x^{n-1} is the least-squares solution.*

Ex. 13.4 *Show that $r^n \cdot p^n = r^n \cdot r^n$, so that*

$$\alpha_n = \frac{r^n \cdot r^n}{p^n \cdot Qp^n}.$$

The proof of Theorem 13.4 uses induction on the number n. Throughout the following exercises assume that the statements in Theorem 13.4 hold for some fixed n with $2 \leq n < J$ and for $j = 1, 2, ..., n-1$. We prove that they hold also for $n+1$ and $j = 1, 2, ..., n$.

Ex. 13.5 *Show that $p^n \cdot Qp^n = r^n \cdot Qp^n$, so that*

$$\alpha_n = \frac{r^n \cdot r^n}{r^n \cdot Qp^n}. \tag{13.10}$$

Hints: Use Equation (13.7) and the induction assumption concerning (3) of the Theorem.

Ex. 13.6 *Show that $r^{n+1} \cdot r^n = 0$. Hint: Use Equations (13.10) and (13.9).*

Ex. 13.7 *Show that $r^{n+1} \cdot r^j = 0$, for $j = 1, ..., n-1$. Hints: Write out r^{n+1} using Equation (13.9) and r^j using Equation (13.7), and use the induction hypotheses.*

Ex. 13.8 *Show that $r^{n+1} \cdot p^j = 0$, for $j = 1, ..., n$. Hints: Use Equations (13.9) and (13.7) and induction assumptions (2) and (3).*

Ex. 13.9 *Show that $p^{n+1} \cdot Qp^j = 0$, for $j = 1, ..., n-1$. Hints: Use Equation (13.9), the previous exercise, and the induction assumptions.*

The final step in the proof is to show that $p^{n+1} \cdot Qp^n = 0$. But this follows immediately from Equation (13.8).

Chapter 14

Operators

14.1 Chapter Summary

In a broad sense, all iterative algorithms generate a sequence $\{x^k\}$ of vectors. The sequence may converge for any starting vector x^0, or may converge only if the x^0 is sufficiently close to a solution. The limit, when it exists, may depend on x^0, and may, or may not, solve the original problem. Convergence to the limit may be slow and the algorithm may need to be accelerated. The algorithm may involve measured data. The limit may be sensitive to noise in the data and the algorithm may need to be regularized to lessen this sensitivity. The algorithm may be quite general, applying to all problems in a broad class, or it may be tailored to the problem at hand. Each step of the algorithm may be costly, but only a few steps may be needed to produce a suitable approximate answer, or, each step may be easily performed, but many such steps are needed. Although convergence of an algorithm is important, theoretically, sometimes in practice only a few iterative steps are used. In this chapter we consider several classes of operators that play important roles in optimization. Up to now we have largely limited our discussion to real vectors and real matrices. In this chapter it is convenient to broaden the discussion to include complex vectors and matrices.

14.2 Operators

For most of the iterative algorithms we shall consider, the iterative step is

$$x^{k+1} = Tx^k,$$

for some operator T. If T is a continuous operator (and it usually is), and the sequence $\{T^k x^0\}$ converges to $\hat{x}$, then $T\hat{x} = \hat{x}$, that is, $\hat{x}$ is a *fixed point* of the operator T. We denote by $\text{Fix}(T)$ the set of fixed points of T. The convergence of the iterative sequence $\{T^k x^0\}$ will depend on the properties of the operator T.

Our approach here will be to identify several classes of operators for which the iterative sequence is known to converge, to examine the convergence theorems that apply to each class, to describe several applied problems that can be solved by iterative means, to present iterative algorithms for solving these problems, and to establish that the operator involved in each of these algorithms is a member of one of the designated classes. We shall be particularly interested in operators that are nonexpansive, firmly nonexpansive, or averaged, in the sense of the two-norm, as well as operators that are strictly contractive or paracontractive for some norm.

14.3 Contraction Operators

Contraction operators are perhaps the best known class of operators associated with iterative algorithms.

14.3.1 Lipschitz-Continuous Operators

Definition 14.1 *An operator T on $\mathbb{C}^J$ is* Lipschitz continuous, *with respect to a vector norm $||\cdot||$, or L-Lipschitz, if there is a positive constant L such that*

$$||Tx - Ty|| \leq L||x - y||,$$

for all x and y in $\mathbb{C}^J$.

For example, if $f : \mathbb{R} \to \mathbb{R}$, and $g(x) = f'(x)$ is differentiable, the Mean Value Theorem tells us that

$$g(b) = g(a) + g'(c)(b - a),$$

for some c between a and b. Therefore,

$$|f'(b) - f'(a)| \leq |f''(c)||b - a|.$$

If $|f''(x)| \leq L$, for all x, then $g(x) = f'(x)$ is L-Lipschitz. More generally, if $f : \mathbb{C}^J \to \mathbb{R}$ is twice differentiable and $\|\nabla^2 f(x)\|_2 \leq L$, for all x, then $T = \nabla f$ is L-Lipschitz, with respect to the Euclidean norm. The two-norm of the Hessian matrix $\nabla^2 f(x)$ is the largest of the absolute values of its eigenvalues.

14.3.2 Nonexpansive Operators

An important special class of Lipschitz-continuous operators are the nonexpansive, or contractive, operators.

Definition 14.2 *If $L = 1$, then T is said to be* nonexpansive *(ne), or a* contraction, *with respect to the given norm. In other words, T is ne for a given norm if, for every x and y, we have*

$$||Tx - Ty|| \leq ||x - y||.$$

Lemma 14.1 *Let $T : \mathbb{C}^J \to \mathbb{C}^J$ be a nonexpansive operator, with respect to the 2-norm. Then the set F of fixed points of T is a convex set.*

Proof: Select two distinct points a and b in F, a scalar α in the open interval $(0,1)$, and let $c = \alpha a + (1-\alpha)b$. We show that $Tc = c$. Note that

$$a - c = \frac{1-\alpha}{\alpha}(c-b).$$

We have

$$\|a-b\|_2 = \|a-Tc+Tc-b\|_2 \leq \|a-Tc\|_2+\|Tc-b\|_2 = \|Ta-Tc\|_2+\|Tc-Tb\|_2$$

$$\leq \|a-c\|_2 + \|c-b\|_2 = \|a-b\|_2;$$

the last equality follows since $a-c$ is a multiple of $(c-b)$. From this, we conclude that

$$\|a - Tc\|_2 = \|a-c\|_2,$$

$$\|Tc - b\|_2 = \|c-b\|_2,$$

and that $a - Tc$ and $Tc - b$ are positive multiples of one another, that is, there is $\beta > 0$ such that

$$a - Tc = \beta(Tc - b),$$

or

$$Tc = \frac{1}{1+\beta}a + \frac{\beta}{1+\beta}b = \gamma a + (1-\gamma)b.$$

Then inserting $c = \alpha a + (1-\alpha)b$ and $Tc = \gamma a + (1-\gamma)b$ into

$$\|Tc - b\|_2 = \|c-b\|_2,$$

we find that $\gamma = \alpha$ and so $Tc = c$. ∎

The reader should note that the proof of the previous lemma depends heavily on the fact that the norm is the two-norm. If x and y are any nonnegative vectors then $\|x+y\|_1 = \|x\|_1 + \|y\|_1$, so the proof would not hold, if, for example, we used the one-norm instead.

We want to find properties of an operator T that guarantee that the sequence of iterates $\{T^k x_0\}$ will converge to a fixed point of T, for any x^0, whenever fixed points exist. Being nonexpansive is not enough; the nonexpansive operator $T = -I$, where $Ix = x$ is the identity operator, has the fixed point $x = 0$, but the sequence $\{T^k x^0\}$ converges only if $x^0 = 0$.

14.3.3 Strict Contractions

One property that guarantees not only that the iterates converge, but that there is a fixed point is the property of being a strict contraction.

Definition 14.3 *An operator T on $\mathbb{C}^J$ is a* strict contraction *(sc), with respect to a vector norm $||\cdot||$, if there is $r \in (0,1)$ such that*

$$||Tx - Ty|| \leq r||x - y||,$$

for all vectors x and y.

For strict contractions, we have the Banach–Picard Theorem [103].

Theorem 14.1 (The Banach–Picard Theorem) *Let T be sc. Then, there is a unique fixed point of T and, for any starting vector x^0, the sequence $\{T^k x^0\}$ converges to the fixed point.*

The key step in the proof is to show that $\{x^k\}$ is a Cauchy sequence, therefore, it has a limit.

Corollary 14.1 *If T^n is a strict contraction, for some positive integer n, then T has a fixed point.*

Proof: Suppose that $T^n\hat{x} = \hat{x}$. Then

$$T^n T\hat{x} = TT^n\hat{x} = T\hat{x},$$

so that both $\hat{x}$ and $T\hat{x}$ are fixed points of T^n. But T^n has a unique fixed point. Therefore, $T\hat{x} = \hat{x}$. ∎

In many of the applications of interest to us, there will be multiple fixed points of T. Therefore, T will not be sc for any vector norm, and the Banach–Picard fixed-point theorem will not apply. We need to consider other classes of operators. These classes of operators will emerge as we investigate the properties of orthogonal-projection operators.

14.3.4 Eventual Strict Contractions

Consider the problem of finding x such that $x = e^{-x}$. We can see from the graphs of $y = x$ and $y = e^{-x}$ that there is a unique solution, which we shall denote by z. It turns out that $z = 0.56714329040978....$ Let us try to find z using the iterative sequence $x_{k+1} = e^{-x_k}$, starting with some real x_0. Note that we always have $x_k > 0$ for $k = 1, 2, ...$, even if $x_0 < 0$. The operator here is $Tx = e^{-x}$, which, for simplicity, we view as an operator on the nonnegative real numbers.

Since the derivative of the function $f(x) = e^{-x}$ is $f'(x) = -e^{-x}$, we have $|f'(x)| \leq 1$, for all nonnegative x, so T is nonexpansive. But we do not have $|f'(x)| \leq r < 1$, for all nonnegative x; therefore, T is not a strict contraction, when considered as an operator on the nonnegative real numbers.

If we choose $x_0 = 0$, then $x_1 = 1$, $x_2 = 0.368$, approximately, and so on. Continuing this iteration a few more times, we find that after about $k = 14$, the value of x_k settles down to 0.567, which is the answer, to three decimal places. The same thing is seen to happen for any positive starting points x_0. It would seem that T has another property, besides being nonexpansive, that is forcing convergence. What is it?

From the fact that $1 - e^{-x} \leq x$, for all real x, with equality if and only if $x = 0$, we can show easily that, for $r = \max\{e^{-x_1}, e^{-x_2}\}$,

$$|z - x_{k+1}| \leq r|z - x_k|,$$

for $k = 3, 4, ...$ Since $r < 1$, it follows, just as in the proof of the Banach–Picard Theorem, that $\{x_k\}$ is a Cauchy sequence and therefore converges. The limit must be a fixed point of T, so the limit must be z.

Although the operator T is not a strict contraction, with respect to the nonnegative numbers, once we begin to calculate the sequence of iterates the operator T effectively becomes a strict contraction, with respect to the vectors of the particular sequence being constructed, and so the sequence converges to a fixed point of T. We cannot conclude from this that T has a unique fixed point, as we can in the case of a strict contraction; we must decide that by other means.

We note in passing that the operator $Tx = e^{-x}$ is *paracontractive*, so that its convergence is also a consequence of the Elsner–Koltracht–Neumann Theorem 14.3, which we discuss later in this chapter.

14.3.5 Instability

Suppose we rewrite the equation $e^{-x} = x$ as $x = -\log x$, and define $Tx = -\log x$, for $x > 0$. Now our iterative scheme becomes $x_{k+1} = Tx_k = -\log x_k$. A few calculations will convince us that the sequence $\{x_k\}$ is diverging away from the correct answer, not converging to it. The lesson here is that we cannot casually reformulate our problem as a fixed-point problem and expect the iterates to converge to the answer. What matters is the behavior of the operator T.

14.4 Orthogonal-Projection Operators

If C is a closed, nonempty convex set in $\mathbb{C}^J$, and x is any vector, then, as we have seen, there is a unique point $P_C x$ in C closest to x, with respect to the 2-norm. This point is called the orthogonal projection of x onto C. If C is a subspace, then we can get an explicit description of $P_C x$ in terms of x; for general convex sets C, however, we will not be able to express $P_C x$ explicitly, and certain approximations will be needed. Orthogonal projection operators are central to our discussion, and, in this overview, we focus on problems involving convex sets, algorithms involving orthogonal projection onto convex sets, and classes of operators derived from properties of orthogonal-projection operators.

14.4.1 Properties of the Operator P_C

Although we usually do not have an explicit expression for $P_C x$, we can, however, characterize $P_C x$ as the unique member of C for which

$$\langle P_C x - x, c - P_C x\rangle \geq 0,$$

for all c in C; see Proposition 4.4.

14.4.2 P_C Is Nonexpansive

It follows from Corollary 4.1 and Cauchy's Inequality that the orthogonal-projection operator $T = P_C$ is nonexpansive, with respect to the Euclidean norm, that is,

$$||P_C x - P_C y||_2 \leq ||x - y||_2,$$

for all x and y. Because the operator P_C has multiple fixed points, P_C cannot be a strict contraction, unless the set C is a singleton set.

14.4.3 P_C Is Firmly Nonexpansive

Definition 14.4 *An operator T is said to be* firmly nonexpansive *(fne) if*

$$\langle Tx - Ty, x - y\rangle \geq ||Tx - Ty||_2^2,$$

for all x and y in $\mathbb{C}^J$.

Lemma 14.2 *An operator $F : \mathbb{C}^J \to \mathbb{C}^J$ is fne if and only if $F = \frac{1}{2}(I+N)$, for some operator N that is ne with respect to the two-norm.*

Proof: Suppose that $F = \frac{1}{2}(I + N)$. We show that F is fne if and only if N is ne in the two-norm. First, we have

$$\langle Fx - Fy, x - y\rangle = \frac{1}{2}\|x - y\|_2^2 + \frac{1}{2}\langle Nx - Ny, x - y\rangle.$$

Also,

$$\left\|\frac{1}{2}(I+N)x - \frac{1}{2}(I+N)y\right\|_2^2 = \frac{1}{4}\|x-y\|^2 + \frac{1}{4}\|Nx-Ny\|^2 + \frac{1}{2}\langle Nx-Ny, x-y\rangle.$$

Therefore,

$$\langle Fx - Fy, x - y\rangle \geq \|Fx - Fy\|_2^2$$

if and only if

$$\|Nx - Ny\|_2^2 \leq \|x - y\|_2^2.$$

∎

Corollary 14.2 *For $m = 1, 2, ..., M$, let $\alpha_m > 0$, with $\sum_{m=1}^{M} \alpha_m = 1$, and let $F_m : \mathbb{C}^J \to \mathbb{C}^J$ be fne. Then the operator*

$$F = \sum_{m=1}^{M} \alpha_m F_m$$

is also fne. In particular, the arithmetic mean of the F_m is fne.

Corollary 14.3 *An operator F is fne if and only if $I - F$ is fne.*

From Equation (4.4), we see that the operator $T = P_C$ is not simply ne, but fne, as well. A good source for more material on these topics is the book by Goebel and Reich [118].

14.4.4 The Search for Other Properties of P_C

The class of nonexpansive operators is too large for our purposes; the operator $Tx = -x$ is nonexpansive, but the sequence $\{T^k x^0\}$ does not converge, in general, even though a fixed point, $x = 0$, exists. The class of firmly nonexpansive operators is too small for our purposes. Although the convergence of the iterative sequence $\{T^k x^0\}$ to a fixed point does hold for firmly nonexpansive T, whenever fixed points exist, the product of two or more fne operators need not be fne; that is, the class of fne operators is not *closed to finite products.* This poses a problem, since, as we shall see, products of orthogonal-projection operators arise in several of the algorithms we wish to consider. We need a class of operators smaller than the ne ones, but larger than the fne ones, closed to finite products, and for which the sequence of iterates $\{T^k x^0\}$ will converge, for any x^0, whenever fixed points exist. The class we shall consider is the class of *averaged* operators. In all discussion of averaged operators the norm will be the two-norm.

14.5 Two Useful Identities

The identities in the next two lemmas relate an arbitrary operator T to its complement, $G = I - T$, where I denotes the identity operator. These identities will allow us to transform properties of T into properties of G that may be easier to work with. A simple calculation is all that is needed to establish the following lemma.

Lemma 14.3 *Let T be an arbitrary operator T on $\mathbb{C}^J$ and $G = I - T$. Then*

$$||x - y||_2^2 - ||Tx - Ty||_2^2 = 2(\langle Gx - Gy, x - y\rangle) - ||Gx - Gy||_2^2. \quad (14.1)$$

Lemma 14.4 *Let T be an arbitrary operator T on $\mathbb{C}^J$ and $G = I - T$. Then*

$$\langle Tx - Ty, x - y\rangle - ||Tx - Ty||_2^2 =$$

$$\langle Gx - Gy, x - y\rangle - ||Gx - Gy||_2^2.$$

Proof: Use the previous lemma. ∎

14.6 Averaged Operators

The term "averaged operator" appears in the work of Baillon, Bruck and Reich [28, 8]. There are several ways to define averaged operators. One way is based on Lemma 14.2.

Definition 14.5 *An operator $T : \mathbb{C}^J \to \mathbb{C}^J$ is* averaged *(av) if there is an operator N that is ne in the two-norm and $\alpha \in (0, 1)$ such that $T = (1 - \alpha)I + \alpha N$. Then we say that T is α-*averaged.

It follows that T is fne if and only if T is α-averaged for $\alpha = \frac{1}{2}$. Every averaged operator is ne, with respect to the two-norm, and every fne operator is av.

We can also describe averaged operators T in terms of the complement operator, $G = I - T$.

Definition 14.6 *An operator G on $\mathbb{C}^J$ is called ν-*inverse strongly monotone *(ν-ism)[119] (also called co-coercive in [86]) if there is $\nu > 0$ such that*

$$\langle Gx - Gy, x - y\rangle \geq \nu||Gx - Gy||_2^2.$$

Lemma 14.5 *An operator T is ne, with respect to the two-norm, if and only if its complement $G = I - T$ is $\frac{1}{2}$-ism, and T is fne if and only if G is 1-ism, and if and only if G is fne. Also, T is ne if and only if $F = (I + T)/2$ is fne. If G is ν-ism and $\gamma > 0$ then the operator γG is $\frac{\nu}{\gamma}$-ism.*

Lemma 14.6 *An operator T is averaged if and only if $G = I - T$ is ν-ism for some $\nu > \frac{1}{2}$. If G is $\frac{1}{2\alpha}$-ism, for some $\alpha \in (0, 1)$, then T is α-av.*

Proof: We assume first that there is $\alpha \in (0, 1)$ and ne operator N such that $T = (1 - \alpha)I + \alpha N$, and so $G = I - T = \alpha(I - N)$. Since N is ne, $I - N$ is $\frac{1}{2}$-ism and $G = \alpha(I - N)$ is $\frac{1}{2\alpha}$-ism. Conversely, assume that G is ν-ism for some $\nu > \frac{1}{2}$. Let $\alpha = \frac{1}{2\nu}$ and write $T = (1 - \alpha)I + \alpha N$ for $N = I - \frac{1}{\alpha}G$. Since $I - N = \frac{1}{\alpha}G$, $I - N$ is $\alpha\nu$-ism. Consequently $I - N$ is $\frac{1}{2}$-ism and N is ne. ∎

An averaged operator is easily constructed from a given operator N that is ne in the two-norm by taking a convex combination of N and the identity I. The beauty of the class of av operators is that it contains many operators, such as P_C, that are not originally defined in this way. As we shall see shortly, finite products of averaged operators are again averaged, so the product of finitely many orthogonal projections is av.

We present now the fundamental properties of averaged operators, in preparation for the proof that the class of averaged operators is closed to finite products.

Note that we can establish that a given operator is av by showing that there is an α in the interval $(0, 1)$ such that the operator

$$\frac{1}{\alpha}(A - (1 - \alpha)I)$$

is ne. Using this approach, we can easily show that if T is sc, then T is av.

Lemma 14.7 *Let $T = (1 - \alpha)A + \alpha N$ for some $\alpha \in (0, 1)$. If A is averaged and N is nonexpansive then T is averaged.*

Proof: Let $A = (1 - \beta)I + \beta M$ for some $\beta \in (0, 1)$ and ne operator M. Let $1 - \gamma = (1 - \alpha)(1 - \beta)$. Then we have

$$T = (1 - \gamma)I + \gamma[(1 - \alpha)\beta\gamma^{-1}M + \alpha\gamma^{-1}N].$$

Since the operator $K = (1 - \alpha)\beta\gamma^{-1}M + \alpha\gamma^{-1}N$ is easily shown to be ne and the convex combination of two ne operators is again ne, T is averaged. ∎

Corollary 14.4 *If A and B are av and α is in the interval $[0, 1]$, then the operator $T = (1 - \alpha)A + \alpha B$ formed by taking the convex combination of A and B is av.*

Corollary 14.5 *Let $T = (1 - \alpha)F + \alpha N$ for some $\alpha \in (0, 1)$. If F is fne and N is ne then T is averaged.*

The orthogonal-projection operators P_H onto hyperplanes $H = H(a, \gamma)$ are sometimes used with *relaxation*, which means that P_H is replaced by the operator

$$T = (1 - \omega)I + \omega P_H,$$

for some ω in the interval $(0, 2)$. Clearly, if ω is in the interval $(0, 1)$, then T is av, by definition, since P_H is ne. We want to show that, even for ω in the interval $[1, 2)$, T is av. To do this, we consider the operator $R_H = 2P_H - I$, which is reflection through H; that is,

$$P_H x = \frac{1}{2}(x + R_H x),$$

for each x.

Lemma 14.8 *The operator $R_H = 2P_H - I$ is an isometry; that is,*

$$||R_H x - R_H y||_2 = ||x - y||_2,$$

for all x and y, so that R_H is ne.

Lemma 14.9 *For $\omega = 1 + \gamma$ in the interval $[1, 2)$, we have*

$$(1 - \omega)I + \omega P_H = \alpha I + (1 - \alpha)R_H,$$

for $\alpha = \frac{1-\gamma}{2}$; therefore, $T = (1 - \omega)I + \omega P_H$ is av.

The product of finitely many ne operators is again ne, while the product of finitely many fne operators, even orthogonal projections, need not be fne. It is a helpful fact that the product of finitely many av operators is again av.

If $A = (1 - \alpha)I + \alpha N$ is averaged and B is averaged then $T = AB$ has the form $T = (1 - \alpha)B + \alpha NB$. Since B is av and NB is ne, it follows from Lemma 14.7 that T is averaged. Summarizing, we have

Proposition 14.1 *If A and B are averaged, then $T = AB$ is averaged.*

14.7 Gradient Operators

Another type of operator that is averaged can be derived from gradient operators. Let $g(x) : \mathbb{C}^J \to \mathbb{C}$ be a differentiable convex function and

$f(x) = \nabla g(x)$ its gradient. If ∇g is nonexpansive, then, according to Theorem 9.20, ∇g is fne. If, for some $L > 0$, ∇g is L-Lipschitz, for the two-norm, that is,

$$||\nabla g(x) - \nabla g(y)||_2 \leq L||x - y||_2,$$

for all x and y, then $\frac{1}{L}\nabla g$ is ne, therefore fne, and the operator $T = I - \gamma\nabla g$ is av, for $0 < \gamma < \frac{2}{L}$. From Corollary 12.1 we know that the operators P_C are actually gradient operators; $P_C x = \nabla g(x)$ for

$$g(x) = \frac{1}{2}(\|x\|_2^2 - \|x - P_C x\|_2^2).$$

14.8 The Krasnosel'skii–Mann–Opial Theorem

For any operator T that is averaged, convergence of the sequence $\{T^k x^0\}$ to a fixed point of T, whenever fixed points of T exist, is guaranteed by the Krasnosel'skii–Mann–Opial (KMO) Theorem [140, 153, 173]. A version of the KMO Theorem 14.2, with variable coefficients, appears in Reich's paper [178].

Theorem 14.2 *Let T be α-averaged, for some $\alpha \in (0, 1)$. Then, for any x^0, the sequence $\{T^k x^0\}$ converges to a fixed point of T, whenever Fix(T) is nonempty.*

Proof: Let z be a fixed point of T. The identity in Equation (14.1) is the key to proving Theorem 14.2.

Using $Tz = z$ and $(I - T)z = 0$ and setting $G = I - T$ we have

$$||z - x^k||_2^2 - ||Tz - x^{k+1}||_2^2 = 2\langle Gz - Gx^k, z - x^k\rangle - ||Gz - Gx^k||_2^2.$$

Since, by Lemma 14.6, G is $\frac{1}{2\alpha}$-ism, we have

$$||z - x^k||_2^2 - ||z - x^{k+1}||_2^2 \geq \left(\frac{1}{\alpha} - 1\right)||x^k - x^{k+1}||_2^2.$$

Consequently the sequence $\{x^k\}$ is bounded, the sequence $\{||z - x^k||_2\}$ is decreasing and the sequence $\{||x^k - x^{k+1}||_2\}$ converges to zero.

Let x^* be a cluster point of $\{x^k\}$. Then we have $Tx^* = x^*$, so we may use x^* in place of the arbitrary fixed point z. It follows then that the sequence $\{||x^* - x^k||_2\}$ is decreasing; since a subsequence converges to zero, the entire sequence converges to zero. The proof is complete. ∎

An operator T is said to be *asymptotically regular* if, for any x, the sequence $\{\|T^k x - T^{k+1}x\|\}$ converges to zero. The proof of the KMO Theorem 14.2 involves showing that any averaged operator is asymptotically regular. In [173] Opial generalizes the KMO Theorem, proving that, if T is nonexpansive and asymptotically regular, then the sequence $\{T^k x\}$ converges to a fixed point of T, whenever fixed points exist, for any x.

Note that, in the KMO Theorem, we assumed that T is α-averaged, so that $G = I - T$ is ν-ism, for some $\nu > \frac{1}{2}$. But we actually used a somewhat weaker condition on G; we required only that

$$\langle Gz - Gx, z - x\rangle \geq \nu\|Gz - Gx\|^2$$

for z such that $Gz = 0$. This weaker property is called *weakly* ν-ism.

14.9 Affine-Linear Operators

It may not always be easy to decide if a given operator is averaged. The class of affine-linear operators provides an interesting illustration of the problem.

The affine operator $Tx = Bx + d$ will be ne, sc, fne, or av precisely when the linear operator given by multiplication by the matrix B is the same.

When B is Hermitian, we can determine if B belongs to these classes by examining its eigenvalues λ:

- B is nonexpansive if and only if $-1 \leq \lambda \leq 1$, for all λ;
- B is averaged if and only if $-1 < \lambda \leq 1$, for all λ;
- B is a strict contraction if and only if $-1 < \lambda < 1$, for all λ;
- B is firmly nonexpansive if and only if $0 \leq \lambda \leq 1$, for all λ.

Affine linear operators T that arise, for instance, in splitting methods for solving systems of linear equations, generally have non-Hermitian linear part B. Deciding if such operators belong to these classes is more difficult. Instead, we can ask if the operator is *paracontractive*, with respect to some norm.

14.10 Paracontractive Operators

By examining the properties of the orthogonal-projection operators P_C, we were led to the useful class of averaged operators. As we shall see shortly, the orthogonal projections also belong to another useful class, the paracontractions.

Definition 14.7 *A continuous operator T is called* paracontractive *(pc), with respect to a given norm, if T has fixed points and, for every fixed point y of T, we have*

$$||Tx - y|| < ||x - y||,$$

unless $Tx = x$.

Paracontractive operators are studied by Censor and Reich in [80].

Proposition 14.2 *The operators $T = P_C$ are paracontractive, with respect to the Euclidean norm.*

Proof: It follows from Cauchy's Inequality that

$$||P_C x - P_C y||_2 \leq ||x - y||_2,$$

with equality if and only if

$$P_C x - P_C y = \alpha(x - y),$$

for some scalar α with $|\alpha| = 1$. But, because

$$0 \leq \langle P_C x - P_C y, x - y \rangle = \alpha ||x - y||_2^2,$$

it follows that $\alpha = 1$, and so

$$P_C x - x = P_C y - y.$$

∎

When we ask if a given operator T is pc, we must specify the norm. We often construct the norm specifically for the operator involved, as we did earlier in our discussion of strict contractions, in Equation (14.4). To illustrate, we consider the case of affine operators.

14.10.1 Linear and Affine Paracontractions

Let the matrix B be diagonalizable and let the columns of V be an eigenvector basis. Then we have $V^{-1}BV = D$, where D is the diagonal matrix having the eigenvalues of B along its diagonal.

Lemma 14.10 *A square matrix B is diagonalizable if all its eigenvalues are distinct.*

Proof: Let B be J by J. Let λ_j be the eigenvalues of B, $Bx^j = \lambda_j x^j$, and $x^j \neq 0$, for $j = 1, ..., J$. Let x^m be the first eigenvector that is in the span of $\{x_j | j = 1, ..., m-1\}$. Then

$$x^m = a_1 x^1 + ... a_{m-1} x^{m-1},$$

for some constants a_j that are not all zero. Multiply both sides by λ_m to get

$$\lambda_m x^m = a_1 \lambda_m x^1 + ... a_{m-1} \lambda_m x^{m-1}.$$

From

$$\lambda_m x^m = A x^m = a_1 \lambda_1 x^1 + ... a_{m-1} \lambda_{m-1} x^{m-1},$$

it follows that

$$a_1(\lambda_m - \lambda_1) x^1 + ... + a_{m-1}(\lambda_m - \lambda_{m-1}) x^{m-1} = 0,$$

from which we can conclude that some x^n in $\{x^1, ..., x^{m-1}\}$ is in the span of the others. This is a contradiction. ∎

We see from this Lemma that almost all square matrices B are diagonalizable. Indeed, all Hermitian B are diagonalizable. If B has real entries, but is not symmetric, then the eigenvalues of B need not be real, and the eigenvectors of B can have non-real entries. Consequently, we must consider B as a linear operator on $\mathbb{C}^J$, if we are to talk about diagonalizability. For example, consider the real matrix

$$B = \begin{bmatrix} 0 & 1 \\ -1 & 0 \end{bmatrix}.$$

Its eigenvalues are $\lambda = i$ and $\lambda = -i$. The corresponding eigenvectors are $(1, i)^T$ and $(1, -i)^T$. The matrix B is then diagonalizable as an operator on C^2, but not as an operator on $\mathbb{R}^2$.

Proposition 14.3 *Let T be an affine-linear operator whose linear part B is diagonalizable, and $|\lambda| < 1$ for all eigenvalues λ of B that are not equal to one. Then the operator T is pc, with respect to the norm given by Equation (14.4).*

Proof: This is Exercise 14.9. ∎

We see from Proposition 14.3 that, for the case of affine operators T whose linear part is not Hermitian, instead of asking if T is av, we can ask if T is pc; since B will almost certainly be diagonalizable, we can answer this question by examining the eigenvalues of B.

Unlike the class of averaged operators, the class of paracontractive operators is not necessarily closed to finite products, unless those factor operators have a common fixed point.

14.10.2 The Elsner–Koltracht–Neumann Theorem

Our interest in paracontractions is due to the Elsner–Koltracht–Neumann (EKN) Theorem [107]:

Theorem 14.3 *Let T be pc with respect to some vector norm. If T has fixed points, then the sequence $\{T^k x^0\}$ converges to a fixed point of T, for all starting vectors x^0.*

We follow the development in [107].

Theorem 14.4 *Suppose that there is a vector norm on $\mathbb{C}^J$, with respect to which each T_i is a pc operator, for $i = 1, ..., I$, and that $F = \cap_{i=1}^I \text{Fix}(T_i)$ is not empty. For $k = 0, 1, ...,$ let $i(k) = k(\text{mod}\, I) + 1$, and $x^{k+1} = T_{i(k)} x^k$. The sequence $\{x^k\}$ converges to a member of F, for every starting vector x^0.*

Proof: Let $y \in F$. Then, for $k = 0, 1, ...,$

$$||x^{k+1} - y|| = ||T_{i(k)} x^k - y|| \leq ||x^k - y||,$$

so that the sequence $\{||x^k - y||\}$ is decreasing; let $d \geq 0$ be its limit. Since the sequence $\{x^k\}$ is bounded, we select an arbitrary cluster point, x^*. Then $d = ||x^* - y||$, from which we can conclude that

$$||T_i x^* - y|| = ||x^* - y||,$$

and $T_i x^* = x^*$, for $i = 1, ..., I$; therefore, $x^* \in F$. Replacing y, an arbitrary member of F, with x^*, we have that $||x^k - x^*||$ is decreasing. But, a subsequence converges to zero, so the whole sequence must converge to zero. This completes the proof. ∎

Corollary 14.6 *If T is pc with respect to some vector norm, and T has fixed points, then the iterative sequence $\{T^k x^0\}$ converges to a fixed point of T, for every starting vector x^0.*

Corollary 14.7 *If $T = T_I T_{I-1} \cdots T_2 T_1$, and $F = \cap_{i=1}^I \text{Fix}\,(T_i)$ is not empty, then $F = \text{Fix}\,(T)$.*

Proof: The sequence $x^{k+1} = T_{i(k)} x^k$ converges to a member of $\text{Fix}\,(T)$, for every x^0. Select x^0 in F. ∎

Corollary 14.8 *The product T of two or more pc operators T_i, $i = 1, ..., I$ is again a pc operator, if $F = \cap_{i=1}^I \text{Fix}\,(T_i)$ is not empty.*

Proof: Suppose that for $T = T_I T_{I-1} \cdots T_2 T_1$, and $y \in F = \text{Fix}(T)$, we have

$$||Tx - y|| = ||x - y||.$$

Then, since

$$||T_I(T_{I-1} \cdots T_1)x - y|| \leq ||T_{I-1} \cdots T_1 x - y|| \leq ...$$

$$\leq ||T_1 x - y|| \leq ||x - y||,$$

it follows that

$$||T_i x - y|| = ||x - y||,$$

and $T_i x = x$, for each i. Therefore, $Tx = x$. ∎

14.11 Matrix Norms

Any matrix can be turned into a vector by vectorization. Therefore, we can define a norm for any matrix A by simply vectorizing the matrix and taking a norm of the resulting vector; the two-norm of the vectorized matrix A is the *Frobenius norm* of the matrix itself, denoted $||A||_F$. The Frobenius norm does have the property

$$||Ax||_2 \leq ||A||_F ||x||_2,$$

known as *submultiplicativity* so that it is compatible with the role of A as a linear transformation, but other norms for matrices may not be compatible with this role for A. For that reason, we consider *compatible* norms on matrices that are induced from norms of the vectors on which the matrices operate.

14.11.1 Induced Matrix Norms

One way to obtain a compatible norm for matrices is through the use of an induced matrix norm.

Definition 14.8 *Let $||x||$ be any norm on $\mathbb{C}^J$, not necessarily the Euclidean norm, $||b||$ any norm on $\mathbb{C}^I$, and A a rectangular I by J matrix. The* induced matrix norm *of A, simply denoted $||A||$, derived from these two vector norms, is the smallest positive constant c such that*

$$||Ax|| \leq c||x||,$$

for all x in $\mathbb{C}^J$. This induced norm can be written as

$$\|A\| = \max_{x \neq 0}\{\|Ax\|/\|x\|\}.$$

We study induced matrix norms in order to measure the distance $\|Ax - Az\|$, relative to the distance $\|x - z\|$:

$$\|Ax - Az\| \leq \|A\| \, \|x - z\|,$$

for all vectors x and z and $\|A\|$ is the smallest number for which this statement can be made.

14.11.2 Condition Number of a Square Matrix

Let S be a square, invertible matrix and z the solution to $Sz = h$. We are concerned with the extent to which the solution changes as the right side, h, changes. Denote by δ_h a small perturbation of h, and by δ_z the solution of $S\delta_z = \delta_h$. Then $S(z + \delta_z) = h + \delta_h$. Applying the compatibility condition $\|Ax\| \leq \|A\|\|x\|$, we get

$$\|\delta_z\| \leq \|S^{-1}\|\|\delta_h\|,$$

and

$$\|z\| \geq \|h\|/\|S\|.$$

Therefore

$$\frac{\|\delta_z\|}{\|z\|} \leq \|S\| \, \|S^{-1}\| \frac{\|\delta_h\|}{\|h\|}.$$

Definition 14.9 *The quantity $c = \|S\|\|S^{-1}\|$ is the* condition number *of S, with respect to the given matrix norm.*

Note that $c \geq 1$: for any nonzero z, we have

$$\|S^{-1}\| \geq \|S^{-1}z\|/\|z\| = \|S^{-1}z\|/\|SS^{-1}z\| \geq 1/\|S\|.$$

When S is Hermitian and positive-definite, the condition number of S, with respect to the matrix norm induced by the Euclidean vector norm, is

$$c = \lambda_{max}(S)/\lambda_{min}(S),$$

the ratio of the largest to the smallest eigenvalues of S.

14.11.3 Some Examples of Induced Matrix Norms

If we choose the two vector norms carefully, then we can get an explicit description of $\|A\|$, but, in general, we cannot.

For example, let $\|x\| = \|x\|_1$ and $\|Ax\| = \|Ax\|_1$ be the one-norms of the vectors x and Ax, where

$$\|x\|_1 = \sum_{j=1}^{J} |x_j|.$$

Lemma 14.11 *The one-norm of A, induced by the one-norms of vectors in both $\mathbb{C}^J$ and $\mathbb{C}^I$, is*

$$\|A\|_1 = \max\Big\{\sum_{i=1}^{I} |A_{ij}|, j = 1, 2, ..., J\Big\}.$$

Proof: Use basic properties of the absolute value to show that

$$\|Ax\|_1 \leq \sum_{j=1}^{J} \Big(\sum_{i=1}^{I} |A_{ij}|\Big)|x_j|.$$

Then let $j = m$ be the index for which the maximum column sum is reached and select $x_j = 0$, for $j \neq m$, and $x_m = 1$. ∎

The *infinity norm* of the vector x is

$$\|x\|_\infty = \max\{|x_j|, j = 1, 2, ..., J\}.$$

Lemma 14.12 *The infinity norm of the matrix A, induced by the infinity norms of vectors in $\mathbb{C}^J$ and $\mathbb{C}^I$, is*

$$\|A\|_\infty = \max\Big\{\sum_{j=1}^{J} |A_{ij}|, i = 1, 2, ..., I\Big\}.$$

The proof is similar to that of the previous lemma.

Lemma 14.13 *Let M be an invertible matrix and $\|x\|$ any vector norm. Define*

$$\|x\|_M = \|Mx\|.$$

Then, for any square matrix S, the matrix norm

$$\|S\|_M = \max_{x \neq 0}\{\|Sx\|_M / \|x\|_M\}$$

is

$$\|S\|_M = \|MSM^{-1}\|.$$

Proof: The proof is left as an exercise. ∎

In [7] Lemma 14.13 is used to prove the following lemma:

Lemma 14.14 *Let S be any square matrix and let $\epsilon > 0$ be given. Then there is an invertible matrix M such that*

$$||S||_M \leq \rho(S) + \epsilon.$$

14.11.4 The Euclidean Norm of a Square Matrix

We shall be particularly interested in the Euclidean norm (or 2-norm) of the square matrix A, denoted by $||A||_2$, which is the induced matrix norm derived from the Euclidean vector norms.

From the definition of the Euclidean norm of A, we know that

$$||A||_2 = \max\{||Ax||_2/||x||_2\},$$

with the maximum over all nonzero vectors x. Since

$$||Ax||_2^2 = x^\dagger A^\dagger Ax,$$

we have

$$||A||_2 = \sqrt{\max\left\{\frac{x^\dagger A^\dagger Ax}{x^\dagger x}\right\}}, \tag{14.2}$$

over all nonzero vectors x.

Proposition 14.4 *The Euclidean norm of a square matrix is*

$$||A||_2 = \sqrt{\rho(A^\dagger A)};$$

that is, the term inside the square-root in Equation (14.2) is the largest eigenvalue of the matrix $A^\dagger A$.

Proof: Let

$$\lambda_1 \geq \lambda_2 \geq ... \geq \lambda_J \geq 0$$

and let $\{u^j,\ j = 1, ..., J\}$ be mutually orthogonal eigenvectors of $A^\dagger A$ with $||u^j||_2 = 1$. Then, for any x, we have

$$x = \sum_{j=1}^{J} [(u^j)^\dagger x]u^j,$$

while

$$A^\dagger Ax = \sum_{j=1}^{J}[(u^j)^\dagger x]A^\dagger Au^j = \sum_{j=1}^{J}\lambda_j[(u^j)^\dagger x]u^j.$$

It follows that

$$\|x\|_2^2 = x^\dagger x = \sum_{j=1}^{J}|(u^j)^\dagger x|^2,$$

and

$$\|Ax\|_2^2 = x^\dagger A^\dagger Ax = \sum_{j=1}^{J}\lambda_j|(u^j)^\dagger x|^2. \tag{14.3}$$

Maximizing $\|Ax\|_2^2/\|x\|_2^2$ over $x \neq 0$ is equivalent to maximizing $\|Ax\|_2^2$, subject to $\|x\|_2^2 = 1$. The right side of Equation (14.3) is then a convex combination of the λ_j, which will have its maximum when only the coefficient of λ_1 is nonzero. ∎

It can be shown that

$$\|A\|_2^2 \leq \|A\|_1\|A\|_\infty;$$

see [59].

If S is not Hermitian, then the Euclidean norm of S cannot be calculated directly from the eigenvalues of S. Take, for example, the square, non-Hermitian matrix

$$S = \begin{bmatrix} i & 2 \\ 0 & i \end{bmatrix},$$

having eigenvalues $\lambda = i$ and $\lambda = i$. The eigenvalues of the Hermitian matrix

$$S^\dagger S = \begin{bmatrix} 1 & -2i \\ 2i & 5 \end{bmatrix}$$

are $\lambda = 3 + 2\sqrt{2}$ and $\lambda = 3 - 2\sqrt{2}$. Therefore, the Euclidean norm of S is

$$\|S\|_2 = \sqrt{3 + 2\sqrt{2}}.$$

Lemma 14.15 *Let T be an affine-linear operator. Then T is a strict contraction if and only if $\|B\|$, the induced matrix norm of B, is less than one.*

Definition 14.10 *The* spectral radius *of a square matrix B, written $\rho(B)$, is the maximum of $|\lambda|$, over all eigenvalues λ of B.*

Since $\rho(B) \leq ||B||$ for every norm on B induced by a vector norm, B is sc implies that $\rho(B) < 1$. When B is Hermitian, the matrix norm of B induced by the Euclidean vector norm is $||B||_2 = \rho(B)$, so if $\rho(B) < 1$, then B is sc with respect to the Euclidean norm.

Let $Tx = Bx + d$ be an affine operator. When B is not Hermitian, it is not as easy to determine if the affine operator T is sc with respect to a given norm. Instead, we often tailor the norm to the operator T. Suppose that B is a diagonalizable matrix, that is, there is a basis for $\mathbb{C}^J$ consisting of eigenvectors of B. Let $\{u^1, ..., u^J\}$ be such a basis, and let $Bu^j = \lambda_j u^j$, for each $j = 1, ..., J$. For each x in $\mathbb{C}^J$, there are unique coefficients a_j so that

$$x = \sum_{j=1}^{J} a_j u^j.$$

Then let

$$||x|| = \sum_{j=1}^{J} |a_j|. \tag{14.4}$$

Lemma 14.16 *The expression $||\cdot||$ in Equation (14.4) defines a norm on $\mathbb{C}^J$. If $\rho(B) < 1$, then the affine operator T is sc, with respect to this norm.*

It is known that, for any square matrix B and any $\epsilon > 0$, there is a vector norm for which the induced matrix norm satisfies $||B|| \leq \rho(B) + \epsilon$. Therefore, if B is an arbitrary square matrix with $\rho(B) < 1$, there is a vector norm with respect to which B is sc.

14.12 Exercises

Ex. 14.1 *Show that a strict contraction can have at most one fixed point.*

Ex. 14.2 *Let T be sc. Show that the sequence $\{T^k x_0\}$ is a Cauchy sequence. Hint: Consider*

$$||x^k - x^{k+n}|| \leq ||x^k - x^{k+1}|| + ... + ||x^{k+n-1} - x^{k+n}||,$$

and use

$$||x^{k+m} - x^{k+m+1}|| \leq r^m ||x^k - x^{k+1}||.$$

Since $\{x^k\}$ *is a Cauchy sequence, it has a limit, say* $\hat{x}$*. Let* $e^k = \hat{x} - x^k$*. Show that* $\{e^k\} \to 0$*, as* $k \to +\infty$*, so that* $\{x^k\} \to \hat{x}$*. Finally, show that* $T\hat{x} = \hat{x}$*.*

Ex. 14.3 *Suppose that we want to solve the equation*

$$x = \frac{1}{2}e^{-x}.$$

Let $Tx = \frac{1}{2}e^{-x}$ *for* x *in* $\mathbb{R}$*. Show that* T *is a strict contraction, when restricted to nonnegative values of* x*, so that, provided we begin with* $x^0 > 0$*, the sequence* $\{x^k = Tx^{k-1}\}$ *converges to the unique solution of the equation. Hint: Use the Mean Value Theorem 9.3.*

Ex. 14.4 *Prove Lemma 14.13.*

Ex. 14.5 *Prove Lemma 14.16.*

Ex. 14.6 *Show that, if the operator* T *is* α*-av and* $1 > \beta > \alpha$*, then* T *is* β*-av.*

Ex. 14.7 *Prove Lemma 14.5.*

Ex. 14.8 *Prove Corollary 14.2.*

Ex. 14.9 *Prove Proposition 14.3.*

Ex. 14.10 *Show that, if* B *is a linear av operator, then* $|\lambda| < 1$ *for all eigenvalues* λ *of* B *that are not equal to one.*

Ex. 14.11 *An operator* $Q : \mathbb{C}^J \to \mathbb{C}^J$ *is said to be* quasi-nonexpansive *(qne) if* Q *has fixed points, and, for every fixed point* z *of* Q *and for every* x*, we have*

$$\|z - x\| \geq \|z - Qx\|.$$

We say that an operator $R : \mathbb{C}^J \to \mathbb{C}^J$ *is* quasi-averaged *if, for some operator* Q *that is qne with respect to the two-norm and for some* α *in the interval* $(0, 1)$*, we have*

$$R = (1 - \alpha)I + \alpha Q.$$

Show that the KMO Theorem 14.2 holds when averaged operators are replaced by quasi-averaged operators.

Chapter 15

Looking Ahead

15.1 Chapter Summary

In this book we have only scratched the surface of optimization; we have ignored entire branches of optimization, such as discrete optimization, combinatorial optimization, stochastic optimization, and many others. The companion volume [64] continues the discussion of optimization, this time focusing on the use of iterative optimization methods in inverse problems. The discussion there begins with constrained optimization and the use of sequential unconstrained iterative algorithms. In this chapter we preview some of the topics treated in greater detail in [64].

15.2 Sequential Unconstrained Minimization

Consider the problem of optimizing a real-valued function f over a subset C of an arbitrary set X. There may well be no simple way to solve this problem and iterative methods may be required. Many well known iterative optimization methods can be described as *sequential optimization* methods. In such methods we replace the original problem with a sequence of simpler optimization problems, obtaining a sequence $\{x^k\}$ of members of the set X. Our hope is that this sequence $\{x^k\}$ will converge to a solution of the original problem, which, of course, will require a topology on X. We may lower our expectations and ask only that the sequence $\{f(x^k)\}$ converge to $d = \inf_{x \in C} f(x)$. Failing that, we may ask only that the sequence $\{f(x^k)\}$ be nonincreasing. One way to design a sequential optimization algorithm is to use auxiliary functions. At the kth step of the iteration we minimize a function

$$G_k(x) = f(x) + g_k(x),$$

to obtain x^k.

In *sequential unconstrained minimization* (SUM) the auxiliary functions $g_k(x)$ are selected to enforce the constraint that x be in C, as in barrier-function methods, or to penalize violations of that constraint, such as in penalty-function methods.

Auxiliary-function (AF) methods closely resemble SUM. In AF methods certain restrictions are placed on the auxiliary functions $g_k(x)$ to control the behavior of the sequence $\{f(x^k)\}$. Even when there are no constraints, the problem of minimizing a real-valued function may require iteration; the formalism of AF minimization can be useful in deriving such iterative algorithms, as well as in proving convergence. As originally formulated, barrier- and penalty-function algorithms are not in the AF class, but can be reformulated as AF algorithms.

In AF methods the auxiliary functions satisfy additional properties that guarantee that the sequence $\{f(x^k)\}$ is nonincreasing. To have the sequence $\{f(x^k)\}$ converging to d we need to impose an additional condition on the $g_k(x)$, the SUMMA condition [56]. The SUMMA condition may seem quite restrictive and ad hoc, and the resulting SUMMA class of algorithms fairly limited, but this is not the case. Many of the best known iterative optimization methods either are in the SUMMA class, or, like the barrier- and penalty-function methods, can be reformulated as SUMMA algorithms.

15.3 Examples of SUM

Barrier-function algorithms and penalty-function algorithms are two of the best known examples of SUM.

15.3.1 Barrier-Function Methods

Suppose that $C \subseteq \mathbb{R}^J$ and $b : C \to \mathbb{R}$ is a barrier function for C, that is, b has the property that $b(x) \to +\infty$ as x approaches the boundary of

C. At the kth step of the iteration we minimize

$$B_k(x) = f(x) + \frac{1}{k}b(x) \tag{15.1}$$

to get x^k. Then each x^k is in C. We want the sequence $\{x^k\}$ to converge to some x^* in the closure of C that solves the original problem. Barrier-function methods are *interior-point methods* because each x^k satisfies the constraints.

For example, suppose that we want to minimize the function $f(x) = f(x_1, x_2) = x_1^2 + x_2^2$, subject to the constraint that $x_1 + x_2 \geq 1$. The constraint is then written $g(x_1, x_2) = 1 - (x_1 + x_2) \leq 0$. We use the logarithmic barrier function $b(x) = -\log(x_1 + x_2 - 1)$. For each positive integer k, the vector $x^k = (x_1^k, x_2^k)$ minimizing the function

$$B_k(x) = x_1^2 + x_2^2 - \frac{1}{k}\log(x_1 + x_2 - 1) = f(x) + \frac{1}{k}b(x)$$

has entries

$$x_1^k = x_2^k = \frac{1}{4} + \frac{1}{4}\sqrt{1 + \frac{4}{k}}.$$

Notice that $x_1^k + x_2^k > 1$, so each x^k satisfies the constraint. As $k \to +\infty$, x^k converges to $(\frac{1}{2}, \frac{1}{2})$, which is the solution to the original problem. The use of the logarithmic barrier function forces $x_1 + x_2 - 1$ to be positive, thereby enforcing the constraint on $x = (x_1, x_2)$.

15.3.2 Penalty-Function Methods

Again, our goal is to minimize a function $f : \mathbb{R}^J \to \mathbb{R}$, subject to the constraint that $x \in C$, where C is a nonempty closed subset of $\mathbb{R}^J$. We select a nonnegative function $p : \mathbb{R}^J \to \mathbb{R}$ with the property that $p(x) = 0$ if and only if x is in C and then, for each positive integer k, we minimize

$$P_k(x) = f(x) + kp(x),$$

to get x^k. We then want the sequence $\{x^k\}$ to converge to some $x^* \in C$ that solves the original problem. In order for this iterative algorithm to be useful, each x^k should be relatively easy to calculate.

If, for example, we should select $p(x) = +\infty$ for x not in C and $p(x) = 0$ for x in C, then minimizing $P_k(x)$ is equivalent to the original problem and we have achieved nothing.

As an example, suppose that we want to minimize the function $f(x) = (x+1)^2$, subject to $x \geq 0$. Let us select $p(x) = x^2$, for $x \leq 0$, and $p(x) = 0$ otherwise. Then $x^k = \frac{-1}{k+1}$, which converges to the right answer, $x^* = 0$, as $k \to \infty$.

15.4 Auxiliary-Function Methods

In this section we define auxiliary-function methods, establish their basic properties, and give several examples.

15.4.1 General AF Methods

Let C be a nonempty subset of an arbitrary set X, and $f : X \to \mathbb{R}$. We want to minimize $f(x)$ over x in C. At the kth step of an auxiliary-function (AF) algorithm we minimize

$$G_k(x) = f(x) + g_k(x)$$

over $x \in C$ to obtain x^k. Our main objective is to select the $g_k(x)$ so that the infinite sequence $\{x^k\}$ generated by our algorithm converges to a solution of the problem; this, of course, requires some topology on the set X. Failing that, we want the sequence $\{f(x^k)\}$ to converge to $d = \inf\{f(x)|x \in C\}$ or, at the very least, for the sequence $\{f(x^k)\}$ to be nonincreasing.

15.4.2 AF Requirements

For AF methods we require that the auxiliary functions $g_k(x)$ be chosen so that $g_k(x) \geq 0$ for all $x \in C$ and $g_k(x^{k-1}) = 0$. We then have the following proposition.

Proposition 15.1 *Let the sequence $\{x^k\}$ be generated by an AF algorithm. Then the sequence $\{f(x^k)\}$ is nonincreasing, and, if d is finite, the sequence $\{g_k(x^k)\}$ converges to zero.*

Proof: We have

$$f(x^k) + g_k(x^k) = G_k(x^k) \leq G_k(x^{k-1}) = f(x^{k-1}) + g_k(x^{k-1}) = f(x^{k-1}).$$

Therefore,

$$f(x^{k-1}) - f(x^k) \geq g_k(x^k) \geq 0.$$

Since the sequence $\{f(x^k)\}$ is decreasing and bounded below by d, the difference sequence must converge to zero, if d is finite; therefore, the sequence $\{g_k(x^k)\}$ converges to zero in this case. ∎

The auxiliary functions used in Equation (15.1) do not have these properties but the barrier-function algorithm can be reformulated as an AF method. The iterate x^k obtained by minimizing $B_k(x)$ in Equation (15.1) also minimizes the function

$$G_k(x) = f(x) + [(k-1)f(x) + b(x)] - [(k-1)f(x^{k-1}) + b(x^{k-1})].$$

The auxiliary functions

$$g_k(x) = [(k-1)f(x) + b(x)] - [(k-1)f(x^{k-1}) + b(x^{k-1})]$$

now have the desired properties. In addition, we have $G_k(x) - G_k(x^k) = g_{k+1}(x)$ for all $x \in C$, which will become significant shortly.

As originally formulated, the penalty-function methods do not fit into the class of AF methods we consider here. However, a reformulation of the penalty-function approach, with $p(x)$ and $f(x)$ switching roles, permits the penalty-function methods to be studied as barrier-function methods, and therefore as acceptable AF methods.

15.5 The SUMMA Class of AF Methods

As we have seen, whenever the sequence $\{x^k\}$ is generated by an AF algorithm, the sequence $\{f(x^k)\}$ is nonincreasing. We want more, however; we want the sequence $\{f(x^k)\}$ to converge to $d = \inf_{x \in C} f(x)$. This happens for those AF algorithms in the SUMMA class [56].

An AF algorithm is said to be in the SUMMA class if the auxiliary functions $g_k(x)$ are chosen so that the SUMMA property holds; that is,

$$G_k(x) - G_k(x^k) \geq g_{k+1}(x) \geq 0, \tag{15.2}$$

for all $x \in C$. As we saw previously, the reformulated barrier-function method is in the SUMMA class. We have the following theorem.

Theorem 15.1 *If the sequence $\{x^k\}$ is generated by an algorithm in the SUMMA class, then the sequence $\{f(x^k)\}$ converges to $d = \inf_{x \in C} f(x)$.*

Proof: Suppose that there is $d^* > d$ with $f(x^k) \geq d^*$, for all k. Then there is z in C with

$$f(x^k) \geq d^* > f(z) \geq d,$$

for all k. Using the inequality (15.2) we have

$$g_k(z) - g_{k+1}(z) \geq f(x^k) + g_k(x^k) - f(z) \geq f(x^k) - f(z) \geq d^* - f(z) > 0.$$

This tells us that the nonnegative sequence $\{g_k(z)\}$ is decreasing, but that successive differences remain bounded away from zero, which cannot happen. ∎

A wide variety of well known iterative optimization algorithms either are in the SUMMA class, or can be reformulated to be in this class. The book [64] uses this fact as a unifying theme, with many of these algorithms discussed in detail, under the umbrella of SUMMA.

Bibliography

[1] Albright, B. (2007) "An introduction to simulated annealing." *The College Mathematics Journal*, **38(1)**, pp. 37–42.

[2] Anderson, A., and Kak, A. (1984) "Simultaneous algebraic reconstruction technique (SART): A superior implementation of the ART algorithm." *Ultrasonic Imaging*, **6**, pp. 81–94.

[3] Attouch, H. (1984) *Variational Convergence for Functions and Operators.* Boston: Pitman Advanced Publishing Program.

[4] Attouch, H., and Wets, R. (1989) "Epigraphical analysis." *Ann. Inst. Poincare: Anal. Nonlineaire*, **6**, pp. 73–100.

[5] Aubin, J.-P., (1993) *Optima and Equilibria: An Introduction to Nonlinear Analysis.* Heidelberg, Germany: Springer-Verlag.

[6] Auslander, A., and Teboulle, M. (2006) "Interior gradient and proximal methods for convex and conic optimization." *SIAM Journal on Optimization*, **16(3)**, pp. 697–725.

[7] Axelsson, O. (1994) *Iterative Solution Methods.* Cambridge, UK: Cambridge University Press.

[8] Baillon, J.-B., Bruck, R.E., and Reich, S. (1978) "On the asymptotic behavior of nonexpansive mappings and semigroups in Banach spaces." *Houston Journal of Mathematics*, **4**, pp. 1–9.

[9] Bauschke, H. (1996) "The approximation of fixed points of compositions of nonexpansive mappings in Hilbert space." *Journal of Mathematical Analysis and Applications*, **202**, pp. 150–159.

[10] Bauschke, H., and Borwein, J. (1993) "On the convergence of von Neumann's alternating projection algorithm for two sets." *Set-Valued Analysis*, **1**, pp. 185–212.

[11] Bauschke, H., and Borwein, J. (1996) "On projection algorithms for solving convex feasibility problems." *SIAM Review*, **38(3)**, pp. 367–426.

[12] Bauschke, H., and Borwein, J. (1997) "Legendre functions and the method of random Bregman projections." *Journal of Convex Analysis*, **4**, pp. 27–67.

[13] Bauschke, H., and Borwein, J. (2001) "Joint and separate convexity of the Bregman distance." In *Inherently Parallel Algorithms in Feasibility and Optimization and their Applications*, edited by D. Butnariu, Y. Censor and S. Reich, pp. 23–36, Studies in Computational Mathematics 8. Amsterdam: Elsevier Publ.

[14] Bauschke, H., and Combettes, P. (2001) "A weak-to-strong convergence principle for Fejér monotone methods in Hilbert spaces." *Mathematics of Operations Research*, **26**, pp. 248–264.

[15] Bauschke, H., and Combettes, P. (2003) "Iterating Bregman retractions." *SIAM Journal on Optimization*, **13**, pp. 1159–1173.

[16] Bauschke, H., Combettes, P., and Noll, D. (2006) "Joint minimization with alternating Bregman proximity operators." *Pacific Journal of Optimization*, **2**, pp. 401–424.

[17] Bauschke, H., and Lewis, A. (2000) "Dykstra's algorithm with Bregman projections: A convergence proof." *Optimization*, **48**, pp. 409–427.

[18] Becker, M., Yang, I., and Lange, K. (1997) "EM algorithms without missing data." *Stat. Methods Med. Res.*, **6**, pp. 38–54.

[19] Bertero, M., and Boccacci, P. (1998) *Introduction to Inverse Problems in Imaging*. Bristol, UK: Institute of Physics Publishing.

[20] Bertsekas, D.P. (1997) "A new class of incremental gradient methods for least squares problems." *SIAM J. Optim.*, **7**, pp. 913–926.

[21] Bertsekas, D., and Tsitsiklis, J. (1989) *Parallel and Distributed Computation: Numerical Methods*. Engelwood Cliffs, NJ: Prentice-Hall.

[22] Bliss, G.A. (1925) *Calculus of Variations*, Carus Mathematical Monographs. Providence, RI: American Mathematical Society.

[23] Borwein, J., and Lewis, A. (2000) *Convex Analysis and Nonlinear Optimization*, Canadian Mathematical Society Books in Mathematics. New York: Springer-Verlag.

[24] Boyd, S., and Vandenberghe, L. (2004) *Convex Optimization*. Cambridge, UK: Cambridge University Press.

[25] Bregman, L.M. (1967) "The relaxation method of finding the common point of convex sets and its application to the solution of problems in convex programming." *USSR Computational Mathematics and Mathematical Physics*, **7**, pp. 200–217.

[26] Bregman, L., Censor, Y., and Reich, S. (1999) "Dykstra's algorithm as the nonlinear extension of Bregman's optimization method." *Journal of Convex Analysis*, **6(2)**, pp. 319–333.

[27] Browne, J., and A. DePierro, A. (1996) "A row-action alternative to the EM algorithm for maximizing likelihoods in emission tomography." *IEEE Trans. Med. Imag.*, **15**, pp. 687–699.

[28] Bruck, R.E., and Reich, S. (1977) "Nonexpansive projections and resolvents of accretive operators in Banach spaces." *Houston Journal of Mathematics*, **3**, pp. 459–470.

[29] Bruckstein, A., Donoho, D., and Elad, M. (2009) "From sparse solutions of systems of equations to sparse modeling of signals and images." *SIAM Review*, **51(1)**, pp. 34–81.

[30] Burden, R.L., and Faires, J.D. (1993) *Numerical Analysis*. Boston: PWS-Kent.

[31] Butnariu, D., Byrne, C., and Censor, Y. (2003) "Redundant axioms in the definition of Bregman functions." *Journal of Convex Analysis*, **10**, pp. 245–254.

[32] Byrne, C., and Fitzgerald, R. (1979) "A unifying model for spectrum estimation." In *Proceedings of the RADC Workshop on Spectrum Estimation*, Griffiss AFB, Rome, NY, October.

[33] Byrne, C., and Fitzgerald, R. (1982) "Reconstruction from partial information, with applications to tomography." *SIAM J. Applied Math.*, **42(4)**, pp. 933–940.

[34] Byrne, C., Fitzgerald, R., Fiddy, M., Hall, T., and Darling, A. (1983) "Image restoration and resolution enhancement." *J. Opt. Soc. Amer.*, **73**, pp. 1481–1487.

[35] Byrne, C., and Fitzgerald, R. (1984) "Spectral estimators that extend the maximum entropy and maximum likelihood methods." *SIAM J. Applied Math.*, **44(2)**, pp. 425–442.

[36] Byrne, C., Levine, B.M., and Dainty, J.C. (1984) "Stable estimation of the probability density function of intensity from photon frequency counts." *JOSA Communications*, **1(11)**, pp. 1132–1135.

[37] Byrne, C., and Fiddy, M. (1987) "Estimation of continuous object distributions from Fourier magnitude measurements." *JOSA A*, **4**, pp. 412–417.

[38] Byrne, C., and Fiddy, M. (1988) "Images as power spectra; reconstruction as Wiener filter approximation." *Inverse Problems*, **4**, pp. 399–409.

[39] Byrne, C. (1993) "Iterative image reconstruction algorithms based on cross-entropy minimization." *IEEE Transactions on Image Processing*, **IP-2**, pp. 96–103.

[40] Byrne, C. (1995) "Erratum and addendum to 'Iterative image reconstruction algorithms based on cross-entropy minimization'." *IEEE Transactions on Image Processing*, **IP-4**, pp. 225–226.

[41] Byrne, C. (1996) "Iterative reconstruction algorithms based on cross-entropy minimization." In *Image Models (and Their Speech Model Cousins)*, S.E. Levinson and L. Shepp, editors, IMA Volumes in Mathematics and Its Applications, Volume 80, pp. 1–11. New York: Springer-Verlag.

[42] Byrne, C. (1996) "Block-iterative methods for image reconstruction from projections." *IEEE Transactions on Image Processing*, **IP-5**, pp. 792–794.

[43] Byrne, C. (1997) "Convergent block-iterative algorithms for image reconstruction from inconsistent data." *IEEE Transactions on Image Processing*, **IP-6**, pp. 1296–1304.

[44] Byrne, C. (1998) "Accelerating the EMML algorithm and related iterative algorithms by rescaled block-iterative (RBI) methods." *IEEE Transactions on Image Processing*, **IP-7**, pp. 100–109.

[45] Byrne, C. (1998) "Iterative algorithms for deblurring and deconvolution with constraints." *Inverse Problems*, **14**, pp. 1455–1467.

[46] Byrne, C. (2000) "Block-iterative interior point optimization methods for image reconstruction from limited data." *Inverse Problems*, **16**, pp. 1405–1419.

[47] Byrne, C. (2001) "Bregman-Legendre multi-distance projection algorithms for convex feasibility and optimization." In *Inherently Parallel Algorithms in Feasibility and Optimization and their Applications*, edited by D. Butnariu, Y. Censor and S. Reich, pp. 87–100, Studies in Computational Mathematics 8. Amsterdam: Elsevier Publ.

[48] Byrne, C. (2001) "Likelihood maximization for list-mode emission tomographic image reconstruction."*IEEE Transactions on Medical Imaging*, **20(10)**, pp. 1084–1092.

[49] Byrne, C., and Censor, Y. (2001) "Proximity function minimization using multiple Bregman projections, with applications to split feasibility and Kullback-Leibler distance minimization." *Annals of Operations Research*, **105**, pp. 77–98.

[50] Byrne, C. (2002) "Iterative oblique projection onto convex sets and the split feasibility problem." *Inverse Problems*, **18**, pp. 441–453.

[51] Byrne, C. (2004) "A unified treatment of some iterative algorithms in signal processing and image reconstruction." *Inverse Problems*, **20**, pp. 103–120.

[52] Byrne, C. (2005) "Choosing parameters in block-iterative or ordered-subset reconstruction algorithms." *IEEE Transactions on Image Processing*, **14(3)**, pp. 321–327.

[53] Byrne, C. (2005) *Signal Processing: A Mathematical Approach.* Wellesley, MA: A K Peters.

[54] Byrne, C., and Ward, S. (2005) "Estimating the largest singular value of a sparse matrix." Unpublished notes.

[55] Byrne, C. (2007) *Applied Iterative Methods.* Wellesley, MA: A K Peters.

[56] Byrne, C. (2008) "Sequential unconstrained minimization algorithms for constrained optimization." *Inverse Problems*, **24(1)**, article no. 015013.

[57] Byrne, C. (2009) "Block-iterative algorithms." *International Transactions in Operations Research*, **16(4)**, pp. 427–463.

[58] Byrne, C. (2009) "Bounds on the largest singular value of a matrix and the convergence of simultaneous and block-iterative algorithms for sparse linear systems." *International Transactions in Operations Research*, **16(4)**, pp. 465–479.

[59] Byrne, C. (2009) *Applied and Computational Linear Algebra: A First Course*, Available as a PDF file at my web site, http://faculty.uml.edu/cbyrne/cbyrne.html.

[60] Byrne, C., and Eggermont, P. (2011) "EM Algorithms." In *Handbook of Mathematical Methods in Imaging*, Otmar Scherzer, ed., pp. 271–344. Heidelberg, Germany: Springer-Science.

[61] Byrne, C., Censor, Y., A. Gibali, A., and Reich, S. (2012) "The split common null point problem." *Journal of Nonlinear and Convex Analysis*, **13**, pp. 759–775.

[62] Byrne, C. (2012) "Alternating minimization as sequential unconstrained minimization: a survey." *Journal of Optimization Theory and Applications*, electronic **154(3)**, DOI 10.1007/s1090134-2, (2012), and hardcopy **156(3)**, pp. 554–566.

[63] Byrne, C. (2014) "An elementary proof of convergence of the forward-backward splitting algorithm." *Journal of Nonlinear and Convex Analysis*, **15(4)**, pp. 681–691.

[64] Byrne, C. (2014) *Iterative Optimization in Inverse Problems.* Boca Raton, FL: CRC Press.

[65] Candès, E., Romberg, J., and Tao, T. (2006) "Robust uncertainty principles: Exact signal reconstruction from highly incomplete frequency information." *IEEE Transactions on Information Theory*, **52(2)**, pp. 489–509.

[66] Candès, E., and Romberg, J. (2007) "Sparsity and incoherence in compressive sampling." *Inverse Problems*, **23(3)**, pp. 969–985.

[67] Candès, E., Wakin, M., and Boyd, S. (2007) "Enhancing sparsity by reweighted l_1 minimization." *J. Fourier Anal. Appl.*, **14**, pp. 877–905.

[68] Carlson, D., Johnson, C., Lay, D., and Porter, A.D. (2002) *Linear Algebra Gems: Assets for Undergraduates*, The Mathematical Society of America, MAA Notes **59**.

[69] Censor, Y., Bortfeld, T., Martin, B., and Trofimov, A. (2006) "A unified approach for inversion problems in intensity-modulated radiation therapy." *Physics in Medicine and Biology*, **51**, pp. 2353–2365.

[70] Censor, Y., Eggermont, P.P.B., and Gordon, D. (1983) "Strong underrelaxation in Kaczmarz's method for inconsistent systems."*Numerische Mathematik*, **41**, pp. 83–92.

[71] Censor, Y., and Elfving, T. (1994) "A multi-projection algorithm using Bregman projections in a product space." *Numerical Algorithms*, **8**, pp. 221–239.

[72] Censor, Y., Elfving, T., Herman, G.T., and Nikazad, T. (2008) "On diagonally-relaxed orthogonal projection methods." *SIAM Journal on Scientific Computation*, **30(1)**, pp. 473–504.

[73] Censor, Y., Elfving, T., Kopf, N., and Bortfeld, T. (2005) "The multiple-sets split feasibility problem and its application for inverse problems." *Inverse Problems*, **21**, pp. 2071–2084.

[74] Censor, Y., Gibali, A., and Reich, S. (2011) "The subgradient extragradient method for solving variational inequalities in Hilbert space." *Journal of Optimization Theory and Applications*, **148**, pp. 318–335.

[75] Censor, Y., Gibali, A., and Reich, S. (2012) "Algorithms for the split variational inequality problem." *Numerical Algorithms*, **59**, pp. 301–323.

[76] Censor, Y., Gordon, D., and Gordon, R. (2001) "Component averaging: An efficient iterative parallel algorithm for large and sparse unstructured problems." *Parallel Computing*, **27**, pp. 777–808.

[77] Censor, Y., Gordon, D., and Gordon, R. (2001) "BICAV: A block-iterative, parallel algorithm for sparse systems with pixel-related weighting." *IEEE Transactions on Medical Imaging*, **20**, pp. 1050–1060.

[78] Censor, Y., Iusem, A., and Zenios, S. (1998) "An interior point method with Bregman functions for the variational inequality problem with paramonotone operators." *Mathematical Programming*, **81**, pp. 373–400.

[79] Censor, Y., and Reich, S. (1998) "The Dykstra algorithm for Bregman projections." *Communications in Applied Analysis*, **2**, pp. 323–339.

[80] Censor, Y., and Reich, S. (1996) "Iterations of paracontractions and firmly nonexpansive operators with applications to feasibility and optimization." *Optimization*, **37**, pp. 323–339.

[81] Censor, Y., and Segman, J. (1987) "On block-iterative maximization." *J. of Information and Optimization Sciences*, **8**, pp. 275–291.

[82] Censor, Y., and Zenios, S.A. (1992) "Proximal minimization algorithm with D-functions." *Journal of Optimization Theory and Applications*, **73(3)**, pp. 451–464.

[83] Censor, Y., and Zenios, S.A. (1997) *Parallel Optimization: Theory, Algorithms and Applications.* New York: Oxford University Press.

[84] Cheney, W., and Goldstein, A. (1959) "Proximity maps for convex sets." *Proc. Amer. Math. Soc.*, **10**, pp. 448–450.

[85] Cimmino, G. (1938) "Calcolo approssimato per soluzioni dei sistemi di equazioni lineari." *La Ricerca Scientifica XVI, Series II, Anno IX*, **1**, pp. 326–333.

[86] Combettes, P. (2001) "Quasi-Fejérian analysis of some optimization algorithms." *Studies in Computational Mathematics*, **8**, pp. 115–152.

[87] Combettes, P. (2001) "Quasi-Fejérian analysis of some optimization algorithms." In *Inherently Parallel Algorithms in Feasibility and Optimization and their Applications*, edited by D. Butnariu, Y. Censor and S. Reich, pp. 87–100, Studies in Computational Mathematics 8. Amsterdam: Elsevier Publ.

[88] Combettes, P., and Wajs, V. (2005) "Signal recovery by proximal forward-backward splitting." *Multiscale Modeling and Simulation*, **4(4)**, pp. 1168–1200.

[89] Conn, A., Scheinberg, K., and Vicente, L. (2009) *Introduction to Derivative-Free Optimization*, MPS-SIAM Series on Optimization. Philadelphia: Society for Industrial and Applied Mathematics.

[90] Csiszár, I. (1975) "I-divergence geometry of probability distributions and minimization problems." *The Annals of Probability*, **3(1)**, pp. 146–158.

[91] Csiszár, I. (1989) "A geometric interpretation of Darroch and Ratcliff's generalized iterative scaling." *The Annals of Statistics*, **17(3)**, pp. 1409–1413.

[92] Csiszár, I., and Tusnády, G. (1984) "Information geometry and alternating minimization procedures." *Statistics and Decisions*, **Supp. 1**, pp. 205–237.

[93] Darroch, J., and Ratcliff, D. (1972) "Generalized iterative scaling for log-linear models." *Annals of Mathematical Statistics*, **43**, pp. 1470–1480.

[94] Dempster, A.P., Laird, N.M., and Rubin, D.B. (1977) "Maximum likelihood from incomplete data via the EM algorithm." *Journal of the Royal Statistical Society, Series B*, **37**, pp. 1–38.

[95] De Pierro, A., and Iusem, A. (1990) "On the asymptotic behavior of some alternate smoothing series expansion iterative methods." *Linear Algebra and Its Applications*, **130**, pp. 3–24.

[96] Deutsch, F., and Yamada, I. (1998) "Minimizing certain convex functions over the intersection of the fixed point sets of non-expansive mappings." *Numerical Functional Analysis and Optimization*, **19**, pp. 33–56.

[97] Dines, K., and Lyttle, R. (1979) "Computerized geophysical tomography." *Proc. IEEE*, **67**, pp. 1065–1073.

[98] Donoho, D. (2006) "Compressed sensing." *IEEE Transactions on Information Theory*, **52(4)**, pp. 1289–1306.

[99] Dorfman, R., Samuelson, P., and Solow, R. (1958) *Linear Programming and Economic Analysis*. New York: McGraw-Hill.

[100] Driscoll, P., and Fox, W. (1996) "Presenting the Kuhn-Tucker conditions using a geometric method." *The College Mathematics Journal*, **38(1)**, pp. 101–108.

[101] Duffin, R., Peterson, E., and Zener, C. (1967) *Geometric Programming: Theory and Applications*. New York: Wiley.

[102] Duda, R., Hart, P., and Stork, D. (2001) *Pattern Classification*. New York: Wiley.

[103] Dugundji, J. (1970) *Topology*. Boston: Allyn and Bacon, Inc.

[104] Dykstra, R. (1983) "An algorithm for restricted least squares regression." *J. Amer. Statist. Assoc.*, **78 (384)**, pp. 837–842.

[105] Eggermont, P.P.B., Herman, G.T., and Lent, A. (1981) "Iterative algorithms for large partitioned linear systems, with applications to image reconstruction." *Linear Algebra and Its Applications*, **40**, pp. 37–67.

[106] Eggermont, P., and LaRiccia, V. (2001) *Maximum Penalized Likelihood Estimation*. New York: Springer.

[107] Elsner, L., Koltracht, L., and Neumann, M. (1992) "Convergence of sequential and asynchronous nonlinear paracontractions." *Numerische Mathematik*, **62**, pp. 305–319.

[108] Facchinei, F., and Pang, J.S. (2003) *Finite Dimensional Variational Inequalities and Complementarity Problems*, Volumes I and II. New York: Springer-Verlag.

[109] Fang, S-C., and Puthenpura, S. (1993) *Linear Optimization and Extensions: Theory and Algorithms*. Englewood Cliffs, NJ: Prentice-Hall.

[110] Farkas, J. (1902) "Über die Theorie der einfachen Ungleichungen." *J. Reine Angew. Math.*, **124**, pp. 1–24.

[111] Farncombe, T. (2000) "Functional dynamic SPECT imaging using a single slow camera rotation." *Ph.D. thesis*, Dept. of Physics, University of British Columbia.

[112] Fiacco, A., and McCormick, G. (1990) *Nonlinear Programming: Sequential Unconstrained Minimization Techniques*, SIAM Classics in Mathematics (reissue). Philadelphia, PA: SIAM.

[113] Fiddy, M. (2008) *Private communication.*

[114] Fleming, W. (1965) *Functions of Several Variables.* Reading, MA: Addison-Wesley.

[115] Gale, D. (1960) *The Theory of Linear Economic Models.* New York: McGraw-Hill.

[116] Geman, S., and Geman, D. (1984) "Stochastic relaxation, Gibbs distributions and the Bayesian restoration of images." *IEEE Transactions on Pattern Analysis and Machine Intelligence*, **PAMI-6**, pp. 721–741.

[117] Gill, P., Murray, W., Saunders, M., Tomlin, J., and Wright, M. (1986) "On projected Newton barrier methods for linear programming and an equivalence to Karmarkar's projective method." *Mathematical Programming,* **36**, pp. 183–209.

[118] Goebel, K., and Reich, S. (1984) *Uniform Convexity, Hyperbolic Geometry, and Nonexpansive Mappings.* New York: Dekker.

[119] Golshtein, E., and Tretyakov, N. (1996) *Modified Lagrangians and Monotone Maps in Optimization.* New York: John Wiley and Sons, Inc.

[120] Gordan, P. (1873) "Über die Auflösungen linearer Gleichungen mit reelen Coefficienten." *Math. Ann.*, **6**, pp. 23–28.

[121] Gordon, R., Bender, R., and Herman, G.T. (1970) "Algebraic reconstruction techniques (ART) for three-dimensional electron microscopy and x-ray photography." *J. Theoret. Biol.*, **29**, pp. 471–481.

[122] Gordon, D., and Gordon, R.(2005) "Component-averaged row projections: A robust block-parallel scheme for sparse linear systems." *SIAM Journal on Scientific Computing,* **27**, pp. 1092–1117.

[123] Gubin, L.G., Polyak, B.T. and Raik, E.V. (1967) "The method of projections for finding the common point of convex sets." *USSR Computational Mathematics and Mathematical Physics*, **7**, pp. 1–24.

[124] Hager, W. (1988) *Applied Numerical Linear Algebra.* Englewood Cliffs, NJ: Prentice Hall.

[125] Hager, B., Clayton, R., Richards, M., Comer, R., and Dziewonsky, A. (1985) "Lower mantle heterogeneity, dynamic typography and the geoid." *Nature*, **313**, pp. 541–545.

[126] Herman, G. T. (1999) *Private communication.*

[127] Herman, G. T., and Meyer, L. (1993) "Algebraic reconstruction techniques can be made computationally efficient." *IEEE Transactions on Medical Imaging*, **12**, pp. 600–609.

[128] Hildreth, C. (1957) "A quadratic programming procedure." *Naval Research Logistics Quarterly*, **4**, pp. 79–85. Erratum, ibid., p. 361.

[129] Hiriart-Urruty, J.-B., and Lemaréchal, C. (2001) *Fundamentals of Convex Analysis*. Berlin: Springer.

[130] Holte, S., Schmidlin, P., Linden, A., Rosenqvist, G. and Eriksson, L. (1990) "Iterative image reconstruction for positron emission tomography: a study of convergence and quantitation problems." *IEEE Transactions on Nuclear Science*, **37**, pp. 629–635.

[131] Hudson, M., Hutton, B., and Larkin, R. (1992) "Accelerated EM reconstruction using ordered subsets." *Journal of Nuclear Medicine*, **33**, p. 960.

[132] Hudson, H.M., and Larkin, R.S. (1994) "Accelerated image reconstruction using ordered subsets of projection data." *IEEE Transactions on Medical Imaging*, **13**, pp. 601–609.

[133] Jiang, M., and Wang, G. (2003) "Convergence studies on iterative algorithms for image reconstruction." *IEEE Transactions on Medical Imaging*, **22(5)**, pp. 569–579.

[134] Kaczmarz, S. (1937) "Angenäherte Auflösung von Systemen linearer Gleichungen." *Bulletin de l'Academie Polonaise des Sciences et Lettres*, **A35**, pp. 355–357.

[135] Kalman, D. (2009) "Leveling with Lagrange: An alternate view of constrained optimization." *Mathematics Magazine*, **82(3)**, pp. 186–196.

[136] Karmarkar, N. (1984) "A new polynomial-time algorithm for linear programming." *Combinatorica*, **4**, pp. 373–395.

[137] Kocay, W., and Kreher, D. (2004) *Graphs, Algorithms, and Optimization*. Boca Raton, FL: CRC Press.

[138] Körner, T. (1996) *The Pleasures of Counting*. Cambridge, UK: Cambridge University Press.

[139] Korpelevich, G. (1976) "The extragradient method for finding saddle points and other problems." *Ekonomika i Matematcheskie Metody* (in Russian), **12**, pp. 747–756.

[140] Krasnosel'skii, M. (1955) "Two observations on the method of sequential approximations." *Uspeki Mathematicheskoi Nauki* (in Russian), **10(1)**, pp. 123–127.

[141] Kuhn, H., and Tucker, A. (eds.) (1956) *Linear Inequalities and Related Systems*. Annals of Mathematical Studies 38. Princeton, NJ: Princeton University Press.

[142] Kullback, S., and Leibler, R. (1951) "On information and sufficiency."*Annals of Mathematical Statistics*, **22**, pp. 79–86.

[143] Lagarias, J., Reeds, J., Wright, M., and Wright, P. (1998) "Convergence properties of the Nelder-Mead simplex method in low dimensions." *SIAM Journal of Optimization*, **9(1)**, pp. 112–147.

[144] Landweber, L. (1951) "An iterative formula for Fredholm integral equations of the first kind."*Amer. J. of Math.*, **73**, pp. 615–624.

[145] Lange, K., and Carson, R. (1984) "EM reconstruction algorithms for emission and transmission tomography." *Journal of Computer Assisted Tomography*, **8**, pp. 306–316.

[146] Lange, K., Bahn, M. and Little, R. (1987) "A theoretical study of some maximum likelihood algorithms for emission and transmission tomography." *IEEE Trans. Med. Imag.*, **MI-6(2)**, pp. 106–114.

[147] Leahy, R. and Byrne, C. (2000) "Guest editorial: Recent development in iterative image reconstruction for PET and SPECT." *IEEE Trans. Med. Imag.*, **19**, pp. 257–260.

[148] Lent, A., and Censor, Y. (1980) "Extensions of Hildreth's row-action method for quadratic programming." *SIAM Journal on Control and Optimization*, **18**, pp. 444–454.

[149] Levy, A. (2009) *The Basics of Practical Optimization*. Philadelphia: SIAM Publications.

[150] Lucet, Y. (2010) "What shape is your conjugate? A survey of computational convex analysis and its applications." *SIAM Review*, **52(3)**, pp. 505–542.

[151] Luenberger, D. (1969) *Optimization by Vector Space Methods*. New York: John Wiley and Sons, Inc.

[152] Luo, Z., Ma, W., So, A., Ye, Y., and Zhang, S. (2010) "Semidefinite relaxation of quadratic optimization problems." *IEEE Signal Processing Magazine*, **27(3)**, pp. 20–34.

[153] Mann, W. (1953) "Mean value methods in iteration." *Proc. Amer. Math. Soc.*, **4**, pp. 506–510.

[154] Marlow, W. (1978) *Mathematics for Operations Research*. New York: John Wiley and Sons. Reissued in 1993 by Dover.

[155] Marzetta, T. (2003) "Reflection coefficient (Schur parameter) representation for convex compact sets in the plane." *IEEE Transactions on Signal Processing*, **51(5)**, pp. 1196–1210.

[156] McKinnon, K. (1998) "Convergence of the Nelder-Mead simplex method to a non-stationary point." *SIAM Journal on Optimization*, **9(1)**, pp. 148–158.

[157] McLachlan, G.J., and Krishnan, T. (1997) *The EM Algorithm and Extensions*. New York: John Wiley and Sons, Inc.

[158] Metropolis, N., Rosenbluth, A., Rosenbluth, M., Teller, A., and Teller, E. (1953) "Equation of state calculations by fast computing machines." *J. Chem. Phys.*, **21**, pp. 1087–1091.

[159] Moreau, J.-J. (1962) "Fonctions convexes duales et points proximaux dans un espace hilbertien." *C.R. Acad. Sci. Paris Sér. A Math.*, **255**, pp. 2897–2899.

[160] Moreau, J.-J. (1963) "Propriétés des applications 'prox'." *C.R. Acad. Sci. Paris Sér. A Math.*, **256**, pp. 1069–1071.

[161] Moreau, J.-J. (1965) "Proximité et dualité dans un espace hilbertien." *Bull. Soc. Math. France*, **93**, pp. 273–299.

[162] Narayanan, M., Byrne, C., and King, M. (2001) "An interior point iterative maximum-likelihood reconstruction algorithm incorporating upper and lower bounds with application to SPECT transmission imaging." *IEEE Transactions on Medical Imaging*, **TMI-20(4)**, pp. 342–353.

[163] Nasar, S. (1998) *A Beautiful Mind*. New York: Touchstone.

[164] Nash, S., and Sofer, A. (1996) *Linear and Nonlinear Programming*. New York: McGraw-Hill.

[165] Nelder, J., and Mead, R. (1965) "A simplex method for function minimization." *Computing Journal*, **7**, pp. 308–313.

[166] Nesterov, Y., and Nemirovski, A. (1994) *Interior-Point Polynomial Algorithms in Convex Programming*, SIAM Studies in Applied Mathematics. Philadelphia, PA: SIAM.

[167] von Neumann, J., and Morgenstern, O. (1944) *Theory of Games and Economic Behavior*. Princeton, NJ: Princeton University Press.

[168] Niven, I. (1981) *Maxima and Minima Without Calculus*. Washington, D.C.: Mathematical Association of America.

[169] Noor, M.A. (1999) "Some algorithms for general monotone mixed variational inequalities." *Mathematical and Computer Modelling*, **29**, pp. 1–9.

[170] Noor, M.A. (2003) "Extragradient methods for pseudomonotone variational inequalities." *Journal of Optimization Theory and Applications*, **117(3)**, pp. 475–488.

[171] Noor, M.A. (2004) "Some developments in general variational inequalities." *Applied Mathematics and Computation*, **152**, pp. 199–277.

[172] Noor, M.A. (2010) "On an implicit method for nonconvex variational inequalities." *Journal of Optimization Theory and Applications*, **147**, pp. 411–417.

[173] Opial, Z. (1967) "Weak convergence of the sequence of successive approximations for nonexpansive mappings." *Bulletin of the American Mathematical Society*, **73**, pp. 591–597.

[174] Ortega, J., and Rheinboldt, W. (2000) *Iterative Solution of Nonlinear Equations in Several Variables*, Classics in Applied Mathematics 30. Philadelphia, PA: SIAM.

[175] Papoulis, A. (1977) *Signal Analysis*. New York: McGraw-Hill.

[176] Peressini, A., Sullivan, F., and Uhl, J. (1988) *The Mathematics of Nonlinear Programming*. New York: Springer-Verlag.

[177] Quinn, F. (2011) "A science-of-learning approach to mathematics education." *Notices of the American Mathematical Society*, **58**, pp. 1264–1275; see also http://www.math.vt.edu/people/quinn/.

[178] Reich, S. (1979) "Weak convergence theorems for nonexpansive mappings in Banach spaces." *Journal of Mathematical Analysis and Applications*, **67**, pp. 274–276.

[179] Reich, S. (1980) "Strong convergence theorems for resolvents of accretive operators in Banach spaces." *Journal of Mathematical Analysis and Applications*, pp. 287–292.

[180] Reich, S. (1996) "A weak convergence theorem for the alternating method with Bregman distances." In *Theory and Applications of Nonlinear Operators*, pp. 313–318. New York: Marcel Dekker.

[181] Renegar, J. (2001) *A Mathematical View of Interior-Point Methods in Convex Optimization*, MPS-SIAM Series on Optimization. Philadelphia, PA: SIAM.

[182] Rockafellar, R. (1970) *Convex Analysis*. Princeton, NJ: Princeton University Press.

[183] Rockmore, A., and Macovski, A. (1976) "A maximum likelihood approach to emission image reconstruction from projections." *IEEE Transactions on Nuclear Science*, **NS-23**, pp. 1428–1432.

[184] Schelling, T. (1980) *The Strategy of Conflict*. Cambridge, MA: Harvard University Press.

[185] Schmidlin, P. (1972) "Iterative separation of sections in tomographic scintigrams." *Nuklearmedizin*, **11**, pp. 1–16.

[186] Schroeder, M. (1991) *Fractals, Chaos, Power Laws*. New York: W. H. Freeman.

[187] Shepp, L., and Vardi, Y. (1982) "Maximum likelihood reconstruction for emission tomography." *IEEE Transactions on Medical Imaging*, **MI-1**, pp. 113–122.

[188] Shermer, M. (2008) "The Doping Dilemma." *Scientific American*, **April 2008**, pp. 82–89.

[189] Shieh, M., Byrne, C., and Fiddy, M. (2006) "Image reconstruction: A unifying model for resolution enhancement and data extrapolation: Tutorial." *Journal of the Optical Society of America, A*, **23(2)**, pp. 258–266.

[190] Shieh, M., Byrne, C., Testorf, M., and Fiddy, M. (2006) "Iterative image reconstruction using prior knowledge." *Journal of the Optical Society of America, A*, **23(6)**, pp. 1292–1300.

[191] Shieh, M., and Byrne, C. (2006) "Image reconstruction from limited Fourier data." *Journal of the Optical Society of America, A*, **23(11)**, pp. 2732–2736.

[192] Simmons, G. (1972) *Differential Equations, with Applications and Historical Notes*. New York: McGraw-Hill.

[193] Stevens, S. (2008) *Games People Play*. A course on DVD available from The Teaching Company, www.TEACH12.com.

[194] Stiemke, E. (1915) "Über positive Lösungen homogener linearer Gleichungen." *Math. Ann*, **76**, pp. 340–342.

[195] Tanabe, K. (1971) "Projection method for solving a singular system of linear equations and its applications."*Numer. Math.*, **17**, pp. 203–214.

[196] Teboulle, M. (1992) "Entropic proximal mappings with applications to nonlinear programming." *Mathematics of Operations Research*, **17(3)**, pp. 670–690.

[197] Tucker, A. (1956) "Dual systems of homogeneous linear relations." In [141], pp. 3–18.

[198] van der Sluis, A. (1969) "Condition numbers and equilibration of matrices." *Numer. Math.*, **14**, pp. 14–23.

[199] van der Sluis, A., and van der Vorst, H.A. (1990) "SIRT- and CG-type methods for the iterative solution of sparse linear least-squares problems." *Linear Algebra and Its Applications*, **130**, pp. 257–302.

[200] Vardi, Y., Shepp, L.A., and Kaufman, L. (1985) "A statistical model for positron emission tomography." *Journal of the American Statistical Association*, **80**, pp. 8–20.

[201] Woeginger, G. (2009) "When Cauchy and Hölder met Minkowski." *Mathematics Magazine*, **82(3)**, pp. 202–207.

[202] Wright, M. (2005) "The interior-point revolution in optimization: History, recent developments, and lasting consequences." *Bulletin (New Series) of the American Mathematical Society*, **42(1)**, pp. 39–56.

[203] Wright, M. (2009) "The dual flow between linear algebra and optimization." Talk given at the History of Numerical Linear Algebra Minisymposium – Part II, SIAM Conference on Applied Linear Algebra, Monterey, CA, October 28.

[204] Yang, Q. (2004) "The relaxed CQ algorithm solving the split feasibility problem." *Inverse Problems*, **20**, pp. 1261–1266.

Index

For Product Safety Concerns and Information please contact our EU
representative GPSR@taylorandfrancis.com Taylor & Francis Verlag GmbH,
Kaufingerstraße 24, 80331 München, Germany

Printed and bound by CPI Group (UK) Ltd, Croydon, CR0 4YY
01/05/2025
01858335-0003